GENETICS, DISABILITY, AND DEAFNESS

GENETICS, DISABILITY, AND DEAFNESS

John Vickrey Van Cleve, Editor

Gallaudet University Press
Washington, D.C.

Gallaudet University Press
Washington, D.C. 20002

http://gupress.gallaudet.edu

© 2004 by Gallaudet University
All rights reserved. Published 2004
Printed in the United States of America

Library of Congress Cataloging-in-Publication Data

Genetics, disability, and deafness / John Vickrey Van Cleve, editor.
    p. cm.
    Includes bibliographical references and index.
    ISBN 1-56368-307-5 (alk. paper)
    1. Deafness-Genetic aspects-Congresses.   I. Van Cleve, John V.
    RF292.G465   2004
    617.8′042-dc22

                                                    2004053314

⊗ The paper used in this publication meets the minimum requirements of
American National Standard for Information Sciences—Permanence of Paper for
Printed Library Materials, ANSI Z39.48–1984.

The photographs on pages 95 and 101 appear courtesy of the Gallaudet University
Archives, Washington, D.C. The photograph on page 97 appears courtesy of the
Gentile family.

CONTENTS

Introduction   vii

Part 1   SCIENCE, CULTURE, AND   1
         HUMAN VARIATION

         The Science of Human Nature and   5
         the Human Nature of Science
                          Louis Menand

         The Cultural Context of Disability   23
                          Nora Groce

Part 2   DEAFNESS AND GENETICS: 31
         A TROUBLED PAST

         The Real "Toll" of A. G. Bell: 35
         Lessons about Eugenics
                          Brian H. Greenwald

         "True Love and Sympathy":   42
         The Deaf-Deaf Marriages Debate in
         Transatlantic Perspective
                          Joseph J. Murray

         Deafness and Eugenics in the Nazi Era   72
                          John S. Schuchman

Part 3   THE SCIENCE OF   79
         GENETIC DEAFNESS

         The Complexity of Hearing Loss from   81
         a Genetics Perspective
                   Orit Dagan and Karen B. Avraham

         The Epidemiology of Hereditary Deafness: 94
         The Impact of Connexion 26 on the
         Size and Structure of the Deaf Community
                   Walter E. Nance

Part 4   THE USES OF GENETIC KNOWLEDGE   107

         Genes for Deafness and the Genetics Program at   111
         Gallaudet University
                   Kathleen S. Arnos and Arti Pandya

         Deaf and Hearing Adults' Attitudes toward   127
         Genetic Testing for Deafness
                   Anna Middleton

         Negotiating (Genetic) Deafness in a   148
         Bedouin Community
                   Shifra Kisch

         Not This Pig: Dignity, Imagination, and   174
         Informed Consent
                   Mark Willis

Part 5   AN ERA DEFINED BY GENOMICS   187

         Frankenstein, Gattaca, and the Quest   189
         for Perfection
                   Christopher Krentz

         Disability, Democracy, and the New Genetics   202
                   Michael Bérubé

         List of Contributors   221

         Index   223

INTRODUCTION

Disability theorists have argued since the late-twentieth century that disability is a social construct and that cultural and political decisions, rather than biological characteristics, restrict their full and complete participation in society. Historian and activist Paul Longmore, for example, has written that "for the overwhelming majority [of disabled persons] prejudice is a far greater problem than any impairment; discrimination is a bigger obstacle than disability."[1] Deaf people also have tried to shift the focus of public discussion from their personal physical attributes to society's response to them. They have argued that deafness is not fundamentally different from ethnicity and should be viewed from the same perspective. The use of accessible language, either a signed language or a spoken language in written form, they continue, renders deafness an interesting human variation, one that should be cherished, respected, even preserved.[2] In these views, the "problems" encountered by people who are deaf or disabled are variable, socially constructed, and not inherent in their biological being.

The attitude of the general, nondisabled public is different. In the popular imagination, disability "promises an unmistakable and noncontingent correspondence between biology and the self," as one author has noted.[3] The even more radical view that biology is destiny is gaining adherents in the United States, fueled in part by the claims of evolutionary psychologists, who view individual realization and social interaction within a framework of supposed evolutionary selection of biologically determined behavioral traits.[4] Louis Menand notes in the first essay in this volume that even such behavioral attributes as anxiety or a taste for novelty recently have been ascribed to the effects of immutable genes rather than to complex human interactions within a specific cultural context. Current social arrangements, cultural habits, and public resource allocations are seen as timeless and essential to national well-being. In this popular view, then, improvement in the conditions of people with disabilities or deafness must come from re-habilitative technology, such as cochlear implants, or from changes in the genes that produce human variability.

The Human Genome Project and other large scientific projects have contributed to the popular interest in genetics and their influence on human variability, behavior, and development. Generally, attempts to find genetic "cures" for disabilities, or other conditions that some peo-ple believe are nonconforming or negative in their effects on human potential, are applauded as progressive wonders of the modern age. Yet their realization in practice has been complex and raises troubling ques-tions, such as when, under which conditions—or whether at all—it is ethical to use in vitro fertilization to select an embryo with particular, desired genes, or to use prenatal diagnoses and abortion to deselect for other, undesirable, biological characteristics.

This volume, drawn from conference papers delivered at Gallaudet University in 2003, addresses these issues by bringing together essays from science and humanism, history and the present, to show the many ways that disability, deafness, and the new genetics can interact and what their interaction means for society. These questions are timely, as prenatal diagnosis of the most frequent form of genetic deafness, for example, is now easily accomplished. Indeed, in her study, "Deaf and Hearing Adults' Attitudes toward Genetic Testing for Deafness," Anna Middleton notes widely reported examples of couples attempting to use genetic knowledge and technology both to select for and against a gene that causes deafness.

The historical sections show how deafness and genetics have been linked in the past and how deaf people have addressed eugenic concerns, but they are also cautionary. The line between scientifically neutral genetics and politically motivated eugenics is neither easily drawn nor easily identified. John Schuchman's essay on deaf people in Nazi Germany recounts what can happen when the state sponsors eugenics programs. Moreover, scientifically validated linkages between human behavior and genes open the doors to genetic manipulation based on blatantly cultural preferences—such as gender, skin color, and height—which are neither consistent across cultures or time nor provably conducive to human well-being.

Several arguments reappear in many of these essays and tie them together into a meaningful whole. One is that "The true good is the different, not the same," as Menand writes in "The Science of Human Nature and the Human Nature of Science." In other words, most authors share the belief that human diversity is a valuable attribute, and they are skeptical of attempts to eliminate it through changes in the human genome. The complexity of even this general observation, however, can be seen in Mark Willis's finely crafted discussion of his reaction to his own genetic makeup, "Not This Pig: Dignity, Imagination, and Informed Consent." On the one hand, Willis sees his hereditary heart problems as a disease. He is a willing participant in studies to find ways to identify and eliminate the genetic abnormality that leads to the premature heart problems he confronts. On the other hand, he writes, "I do not experience vision loss as a disease. It is a different way of perceiving the world, and it is rich with its own sensory skills and sweet satisfactions." Thus, he refuses to be involved, to give his "informed consent," for genetic studies of the particular kind of blindness he has experienced.

The variability of disability—in physical expression, in cultural meaning, and as lived experience—leads to the most important argument to recur in nearly every essay: Choice and democratic values should control social approaches to disability and to the uses of genetic science and technology. The scientists, particularly Walter Nance, the most well-known student of deafness and heredity, tend to emphasize the role of personal choice and the importance of science in offering individuals knowledge from which they can fashion their own decisions about genetics and disability or deafness. Nance writes that "as long as decisions about issues such as marital choices or whether or not to use specific genetic technologies are made by fully competent, fully informed individuals, I'm willing to live with the consequences." Michael Bérubé, by contrast, focuses more on public choices and public policy, arguing that "disability is always and everywhere a public issue." Despite the different focus on where choice should be applied in considerations of genetics and disability, however, both the scientists and the humanists writing for this volume would agree with Bérubé's concluding essay when he argues that disability should be considered "democratically." They also would agree, as emphasized by authors of the essays in the section titled "The Uses of Genetic Knowledge," that any "democratic deliberation" about disability requires "the voices of people with disabilities" to be involved.

These essays are offered, then, as a way to provide context and meaning to a public discussion of difference—its past, how it should be dealt with in the future, and what the role of genetic counseling and genetic manipulation might be as society thinks about disability and deafness and the public and private choices that need to be made in this age of genomics.

ACKNOWLEDGMENTS

Academic conferences require planning, creativity, attention to detail, and money. We were fortunate to have all these in abundance to put on "Genetics, Disability, and Deafness" under the auspices of the Gallaudet University Press Institute in the spring of 2003. I want to thank the other members of the conference planning committee—David Armstrong, Cathy Arnos, Derek Braun, Wendy Grande, Michael Karchmer, and Dan Wallace—for their commitment of time, creativity and good company. Some financial support was provided by the Dean of the Graduate School and Professional Studies at Gallaudet University, Tom Allen, and the members of the Gallaudet University Schaefer Professorship Committee. Paul Kelly, the Gallaudet University Vice President for Administration and Finance, provides the overall financial support necessary for the Gallaudet University Press Institute's operations and for the underwriting of the conferences it sponsors. Without his commitment, this conference could not have taken place. Finally, although I have thanked her once, Wendy Grande deserves special recognition for her unfailing efforts to ensure that the conference logistics were managed flawlessly, the presenters' needs were met, and the attendees derived maximum benefit from the conference.

NOTES

1. Paul Longmore, *Why I Burned My Book and Other Essays on Disability* (Philadelphia: Temple University Press, 2003), p. 130.

2. The classic statement of this position is a book by Carol Padden and Tom Humphries, *Deaf in America: Voices from a Culture* (Cambridge: Harvard University Press, 1988).

3. Natalie A. Dykstra, "'Trying to Idle': Work and Disability in the *Diary of Alice James*," in *The New Disability History: American Perspectives*, ed. Paul K. Longmore and Lauri Umansky (New York: New York University Press, 2001), p. 109.

4. See, for instance, Steven Pinker, *The Blank Slate: The Modern Denial of Human Nature* (New York: Viking, 2002).

PART 1
_____

SCIENCE, CULTURE, AND HUMAN VARIATION

INTRODUCTION

Louis Menand and Nora Groce, the authors of the first two essays in this book, are both well known—Menand for *The Metaphysical Club* and Groce for *Everyone Here Spoke Sign Language: Hereditary Deafness on Martha's Vineyard*.[1] Despite profound differences in their approaches to matters of genetics, culture, and human variation, Menand and Groce suggest a similar need for skepticism and critical thinking in the consideration of human differences and genetics. By implication, both also question the emphasis that American society places on biological solutions—changing human genes—to a range of social problems supposedly caused by human differences. This critique has two major aspects: First, variation is itself a social good, that is, striving to achieve a physical or cultural "norm" for everyone is a mistake; second, cultural, social, or political measures are the appropriate ways to enhance the lives of people with conditions labeled as disabilities.

Menand's Pulitzer Prize–winning *The Metaphysical Club* focuses on a group of American pragmatists and their intellectual milieu during the late-nineteenth and early-twentieth centuries. This was the heyday of eugenics' practices and popularity in the United States. Darwin's theory of evolution had spawned speculation about linkages between genetic characteristics and human behavior; states were beginning to pass laws providing for the sterilization of "genetic inferiors"; new immigration policies overtly favored supposedly superior genetic stocks of some nations over others; Alexander Graham Bell raised the fear of the growth of a "deaf variety of the human race"; and the Nazis had not yet discredited eugenics through their program of sterilizations, forced abortions, and murders of persons with alleged genetic defects.

Menand uses some of the material from this text and the period he knows well to develop his arguments in the essay, "The Science of Human Nature and the Human Nature of Science." Therein Menand states flatly, "There is a great battle going on in our intellectual culture today . . . between people who believe that science opens new possibilities for human life and people who worry that it closes them." Menand is in the second camp, concerned that a focus on the genome limits human potential. The people who advance genetic explanations of behavior, he writes, "speak as though genes are some sort of irreducible reality, as though they are a power behind human affairs that cannot be abrogated or countervailed against." He uses both contemporary and historical arguments, tracing the thinking of individuals such as William James, Horace Kallen, and Alain Locke to challenge biological determinism and argue that human intelligence has "released us from the prison of biology."

Groce's *Everyone Here Spoke Sign Language* is a classic in comparative anthropology, arguing that a society's reaction to deafness, and by implication other characteristics or disabilities, is variable, depending on cultural considerations. Groce has studied disability in many different cultures, and in her essay, "The Cultural Context of Disability," she reviews the ways that cultures have defined and interpreted disability and creates an interpretive framework for discussing the relationship between culture and disability. She makes two core arguments. The first is that the lives of individuals with disabilities "will in large measure be determined not by the fact that they have a disability but by the way the society in which they live conceptualizes what it means to be 'disabled.'" The second is that society's reaction to disability is culturally dependent; that is, societies do not, in fact, all conceptualize all disabilities the same way. In other words, society and culture can be constructed in such a way that people with a variety of physical and mental attributes can lead fulfilling and useful lives.

NOTE

1. Louis Menand, *The Metaphysical Club: A Story of Ideas in America* (New York: Farrar, Straus and Giroux, 2001); Nora Groce, *Everyone Here Spoke Sign Language: Hereditary Deafness on Martha's Vineyard* (Cambridge, Mass.: Harvard University Press, 1985).

THE SCIENCE OF HUMAN NATURE AND THE HUMAN NATURE OF SCIENCE

Louis Menand

In 1889, the German biologist August Weissmann showed that mice whose tails are cut off do not produce short-tailed offspring. It was a step forward for science, but a step backward for civilization. Weissmann's discovery was good for science because, contrary to what many scientists had believed, acquired characteristics are not, of course, heritable. Weissmann's experiment closed the book on the neo-Lamarckianism that many scientists had adopted in order to blunt the edge on Darwin's theory of natural selection. Darwin had left evolution to chance, and had pretty much ruled out the inevitability, if not the possibility, of progress. "Never use the word[s] higher & lower," he had once written in a note to himself.[1] But if there was something human beings could do to affect the course of evolutionary development, then the story of evolution might have a happy ending after all. If we just

---

Parts of this essay are drawn from my book *The Metaphysical Club* and from articles by me published in the *New Yorker*.

kept cutting those nasty-looking tails down, generation after generation, we might eventually get rid of them altogether.

Weissmann's demonstration was bad for civilization, therefore, for precisely the same reason that it was good for science. If characters are fixed, if the genome is hermetically sealed off from the environment, then biological characteristics must be immutable. We think of Lamarckianism as retrograde, but, in a century in which white Americans and Europeans believed, almost universally, that the human races are ranked hierarchically, from higher to lower, Lamarckianism held out the hope that with dedicated exposure to Christian civilization, the lesser races might someday be raised up. If enlightenment is not heritable, though, if the tails have to be cut off again in every generation, there was no hope.

Soon after Weissmann announced his conclusions, a huge wave of immigration pounded the United States. Between 1901 and 1910, 8.8 million immigrants were admitted to the United States; 70 percent were from Southern and Eastern Europe, principally Catholics and Jews. Between 1911 and 1920, another 5.7 million people came from abroad, 59 percent of them from Southern and Eastern Europe. By 1910, 40 percent of the population of New York City was foreign-born. At a time when nationality was defined racially, and race was conceived hierarchically, there was widespread anxiety that the presence of large numbers of non-Anglo-Saxon peoples would lead to national degeneration. This is why, in the early years of the twentieth century, the doctrine of eugenicism was not limited to rabble-rousers and bigots. Many of the most educated and progressive thinkers of the time were dedicated eugenicists: Theodore Roosevelt; Oliver Wendell Holmes, Jr.; the sociologist Edward A. Ross; the political scientist Harold Laski; David Starr Jordan, the president of Stanford; Charles William Eliot, the president of Harvard; and even the Marxist revolutionary Emma Goldman.

At the height of the panic over immigration in the United States, there were several efforts to buck the tide of nativist sentiment and assert the virtues of ethnic pluralism. Some of these texts are read today as brave prolepses of multiculturalism—for example, Horace Kallen's anti-anti-immigrant essay, "Democracy Versus the Melting Pot," which he published in 1915. Kallen's essay was written as a response to the eugenicist Edward Ross, who had just published a collection of essays opposing immigration. But in fact, Kallen's science was exactly the same as Ross's. Kallen was a recovered Jew—that is, he had lapsed from the faith of his father, a German immigrant who had become a Boston rabbi, but had rediscovered his Jewishness as a student at Harvard, under the influence of a professor who persuaded him that the Puritans had Hebrew blood. This reconversion inspired Kallen to his core belief: fulfillment in life is a function of cultural identity; cultural identity is a function of ethnicity; and ethnicity is immutable. The most famous line in Kallen's essay on "Democracy Versus the Melting Pot" is this one: "Men may change their clothes, their politics, their wives, their religions, their philosophies, to a greater or lesser extent; they cannot change their grandfathers."[2] Or as he put it again a few years later: "An Irishman is always an Irishman, a Jew always a Jew. . . . Irishman and Jew are facts in nature; citizen and church-member artefacts in civilization."[3]

Kallen thought that racial ancestry was an unalterable constituent of selfhood. This meant, as he put it, that the happiness people pursue in their lives "has its form implied in ancestral endowment."[4] Your hopes and fears, your limitations and your potential, are already there at birth, in your genes. Kallen did not believe that the races (or nationalities, which he defined racially) were equal in natural endowment. He thought that some races were higher than others, and that each race, or ethnic group, had its own inherent characteristics. Nor did he believe in race-mixing, ethnic interbreeding, although he thought that since people generally prefer their own kind, this was not as great a danger as people like Edward Ross feared. Kallen only believed that each ethnic group deserved equal respect—so long as it kept its social place. He espoused a noninvidious form of ethnic and racial segregation: separate but equal, equal as long as separate, higher distinct from lower. Science had let him see no alternative.

There is a great battle going on in our intellectual culture today. It is a battle between people who believe that science opens new possibilities for human life and people who worry that it closes them. I spend most of my time around the second kind of people, people who regard science as more or less an agent of social control disguised as a neutral observer. These people think that scientists are reductive, and that they are too quick to leap from data to prescription. More than this, they think that science refuses to admit the reality of anything that it cannot measure. The great web of metaphor and image that people use to describe and make sense of their experience in the world is dismissed by scientists, who prefer to talk about things like genes and neurotransmitters instead. But to the skeptics, genes and neurotransmitters are just as much imaginary constructs as witches' curses and the Oedipal complex, and a lot less suggestive. They seem hopelessly underpowered explanatory devices, boxed in by the dogmatics of empiricism.

The type of racial science that informed the thinking of people like Ross and Kallen is now, thankfully, discredited. No one believes that the nonheritability of artificially shortened mouse tails has implications for immigration policy. Contemporary anti-immigrationists rely on different arguments. But there are new developments in science today that, in their popularized form, seem to many people to propose restrictions on human possibility. These are behavioral genetics and evolutionary psychology, sometimes referred to as "the new sciences of human nature."

In 1996, scientists announced the discovery of the gene for anxiety. It made the front page of the *New York Times*, which reported that people who are fretful, anxious, and neurotic—"kvetches," in the *Times*'s helpful translation into regional dialect—have a shorter version of a certain gene (*SLC6A4*, which is the serotonin transporter gene on chromosome 17q12, if you want to look it up) than do people of sunny disposition. The only sensible conclusion to be drawn from this report was that if you're a worrier, it is not because you have a lot of things in your life to be worried about. It's just because you're a worrier. Does this make people who worry feel better? Of course not. They *can't* feel better. That's the whole point.

The discovery of the worry gene followed closely on the discovery, reported earlier that year, of the gene associated with the taste for novelty and excitement, the so-called "gene for bungee-jumping." Now, it's easy to make fun of these genes for behaviors, like bungee-jumping, which were not even invented back when our species emerged from the protozoan slime. But the truth is that they are immensely clarifying contributions to thought, for they reverse the usual relation between accident and necessity in human life. For people who lived in New York City, the obvious question raised by the discovery of the gene for anxiety was, if crabby and kvetchy behavior is genetically determined, why does New York City have so much of it? The answer must be that New Yorkers are not neurotic because they live in New York. They live in New York because they're neurotic.

Once this principle has been grasped, many prejudices fall away. Cabdrivers are not impatient because they're cabdrivers; they're cabdrivers because they're impatient. Taking care of small children does not make people feel stressed out; feeling stressed out is what makes people take care of small children. And so on. This is determinism, but it is determinism of a deeply appealing kind. We used to think of our moods and tastes as byproducts of the social and personal relations in which we happened to be stuck. Now we can see that the social and personal relations we are stuck in are only the accidental consequences of our tastes and moods. Many are born impatient; the lucky ones become cabdrivers. The animus seeks its animal. The circularity takes your breath away. It is as though scientists were to explain the behavior of the mosquito by showing it to have a gene for being annoying.

Still, every simplification sooner or later leads to complications, and genetic explanations for human personality and behavior, as delightfully shorn of metaphysical distractions as they seem to be, are no exception. First of all, there is, when you are explaining the basis of personality, the question of what a personality really is. Peter Kramer, in *Listening to Prozac*, a more thoughtful book than it got credit for being when it came out, noted the odd phenomenon of patients who told him, after they were on an anti-depressant medication, that they finally felt "like themselves." For, of course, you can deal with the side effects of having a shortened serotonin transporter gene—the anxiety gene—by taking a selective serotonin reuptake inhibitor, or SSRI. This makes you lose the sensation of anxiety, and the result, even if you have always been an anxious person, can indeed be to make you feel "like yourself." Kramer noted that this was a bit of a paradox: how is it that the chemically altered self feels more real, more genuine, than the biologically natural self?

But it is a paradox only if you assume that the self, or personality, is a stable entity to begin with. This is not a topic to be approached glibly. The subject of identity is a sea on which many philosophers have lost their way. Consider a bundle of sticks from which one stick is removed, then another, and then another. After the removal of which stick does the bundle of sticks cease to be a bundle of sticks? Or take the case of a knife whose blade has been replaced once and whose handle has been replaced three times. What grounds do we have for calling it "the same knife?" As with things, so with selves. The eighteen-year-old who is ready for anything, the puller of all-nighters, the consumer of three pizzas and a six-pack, on the one hand, and the sagging commuter of twenty years later, who staggers home hoping only to have the stamina to make it through the first half of Charlie Rose, on the other, are nominally the same person. But by virtue of what? Of having the same Social Security number? If we have a self, it is never the same self for very long. Identity is the artificial flower on the compost heap of time.

So that when we begin to talk about the self on antidepressant medication, we are immediately aware of a certain fugitive quality in the object of our attention, and chemistry need have nothing to do with it. Mood transformers have many agents. The person who has just polished off an entire pint of coffee ice cream with cookie dough is not the same person who opened the refrigerator door fifteen minutes earlier. The person who spent the night cleaning up after a six-year-old with a stomach virus is not the calm and obliging person who went to bed the night before. The person who paid ten dollars to sit through *Gangs of New York* is not the person who thought this sounded like a really good movie. Probably the only thing to say about our "real" self is that it is the self we are least embarrassed about owning up to. The rest of the time, we're, well, just not ourselves.

In short, if genetic behaviorism is a kind of determinism, it is a very indeterminate kind of determinism. And the reason is that all behavior within the normal range is overdetermined anyway. There are too many inputs for us to be able to distinguish the true cause from what we might call, on an intelligence analogy, causal noise. If one gene is telling you to bite your fingernails, seven other genes might be pumping out the neurological equivalent of Easy Listening music, telling you to lay back and chill out. There are, as well, the environmental triggers, the mental history of the organism, the Rocky Road ice cream, the Prozac, the anhedonic effect of Charlie Rose, and all the rest. Another way to put it is to say that gene-based explanations for human behavior belong to a kind of polytheistic view of the universe. If you think about it, there is not that much difference between saying, "He jumped off the bridge because the gods made him crazy," and saying, "He jumped because his dopamine made him do it." Genetic explanation is a way of ascribing personality and behavior to some involuntary cause. In the ancient world, there were many gods, and if one god put a spell on you—if one god made you a bungee-jumper, let's say—there were plenty of other gods around who could take the spell off. And so it is with genes and medications: few powers are so great that we cannot summon other powers to thwart their effects.

It is interesting that just as genetic explanations for behavior are becoming popularized, there is a parallel fascination with cultural explanations for behavior. In between the old polytheism of the ancient world and the new polytheism of the genome project, there were, of course, a number of fairly successful monotheisms, single-variable explanations that won wide adherence. In between "Circe put a spell on her" and "Her genes made her do it," there were, besides the major religious monotheisms, "her subconscious rage against her parents made her do it," "the iron law of history made her do it," and "market forces made her do it." Those monotheisms are still around, but cultural explanations are by definition polytheistic; they are the "his epistemology made him do it" explanations. The idea is that personality and behavior are determined by cultural input, which is why different cultures produce different sorts of human beings: some are warlike, some are peaceful, some are sexist, some are androgynous, and so on. There is a basic contradiction between the gene-based polytheisms and the cultural polytheisms, since what is wired into the hardware cannot be reconfigured by the software. You can't believe that certain people have a naturally selected gene for aggression and, at the same time, that people become violent by watching *The Terminator*. Most people are likely to want to believe a little of both—that people are what they are, and that they might be made better by going to church more regularly. But one group of gods has got to go. It's like having the Greek gods and the Incan gods in the same pantheon.

What is exasperating to the nonscientific mind about genetic explanations is that the people who advance them speak as though genes are some sort of irreducible reality, as though they are a power behind human affairs that cannot be abrogated or countervailed against. It is a version of the early-twentieth-century belief that since the genome is sealed off from environmental effects, people cannot fundamentally be changed. This is sometimes the claim of evolutionary psychologists: they write as though biology is fate. The trouble with evolutionary psychology seems to me to be that it is not really psychology. Take, for example, two recent works that study the effects of parenting on children's personalities: Judith Harris's *The Nurture Assumption* and Steven Pinker's *The Blank Slate*. The authors claim that shared family environment—that is, parents—have little or no influence on a child's personality. (Strictly speaking, they claim that parenting does not account for the variation in differences in personality, which is what statistical science measures.) They point out that biological siblings who have been reared together are not more alike, or less different, than biological siblings who have been reared in separate families. They conclude that half of personality is the product of what they call the unique environment—that is, the child's personal history—and half is the product of genes. The effect of parenting is statistically insignificant. Harris's argument is that children's peers are the principal source of the nongenetic input. Parents used to obsess about reading bedtime stories regularly to their children. These new sciences of human nature have established that a belief in the civilizing effects of bedtime stories on a child's personality is a modern superstition.

What *is* personality to people like Harris and Pinker, though? The answer is OCEAN: the personality attributes in the Five Factor Model. In this model, personality has exactly five dimensions: people are, in varying degrees, either open to experience or incurious; conscientious or undirected; extroverted or introverted; agreeable or antagonistic; neurotic or stable. OCEAN is the acronym for these spectra. There is no need for finer tuning, because OCEAN accounts for everything. As Pinker puts it, "Most of the 18,000 adjectives of personality traits in an unabridged dictionary can be tied to one of the five dimensions."[5]

What the genetic claims about parenting boil down to, therefore, is that parents cannot turn a fretful child into a serene adult. But parents can make their children into opera buffs, water-skiers, painters, food connoisseurs, bilingual speakers, trumpet players, and churchgoers. Parents introduce their children to the whole supra-biological realm. The claim that chronic anxiety is biological is proven by the fact that an SSRI can relieve it. But that's just the *biology*. The *psychology* is everything that the organism does to cope with its biology. Anxious people develop all kinds of strategies for overcoming, disguising, avoiding, repressing, and even exploiting their tendency to nervousness. I know, because I am someone who, for most of his life, has disguised an inveterate anxiety with an affect of coolness. No one ever reads me as an anxious person, and seeing myself in that mirror helps me to manage my anxiety. Am I therefore an anxious person or a calm person? Strategies like these are acquired—people aren't born with them—and they are constructed from the elements the environment provides. The mind can work only with what it knows, and one of the things it knows are parents, who often become major players in the psychic drama of anxiety maintenance. The mere fact of having the gene for anxiety determines nothing, which is why some anxious people become water-skiers, some become opera buffs, and some are most comfortable speaking in front of large groups. Some anxious people, it's true, sit and stare out the window, brooding on the fact that their parents did not read them enough bedtime stories. These people are unlikely to be relieved by learning that genetic science has determined that bedtime stories are overrated.

The most unfortunate aspect of contemporary evolutionary psychology, in its popularized form, is the obsession with the mean point of the normal distribution. Evolutionary psychologists seem to forget that the mean is a mathematical construct, corresponding to no actual human being. It represents, in many cases, not the acme of attainment, but, on the contrary, the lowest common denominator. But it is often treated as though it were some sort of species norm, the bull's-eye at which civilization aims. The classic case of this kind of apotheosis of the average is the study that discovers the ideal female face by blending all the features people identify as most beautiful. The result is a homogenized, anodyne image with very little aesthetic or erotic appeal. This is because people don't go for faces that deviate from the ideal because they can't have the ideal. The deviation is precisely what makes those faces attractive.

And so it is with most of the things we care about in life: food, friends, recreation, art. Biology reverts to the mean; civilization does not. The mind is a fabulator. It is designed—by natural selection, if you like—to dream up ideas and experiences away from the mean. Its instinct is to be counter-instinctual; otherwise, we could put consciousness to sleep at an early age. The mind has no steady state. It is never satisfied. It induces the organism to go to fantastic lengths to develop capacities that have no biological necessity, and that in some cases, such as bungee-jumping, are completely counter-indicated by

biological conditioning. The more defiant something is of the instinc-
tual and the habitual, the more highly civilization prizes it. This is why
we have the *Guinness Book of World Records*, the Gautama Buddha, and
the Museum of Modern Art. They represent the repudiation of the
norm. The composite beautiful face tells us as much about beauty as a
dish containing all the flavors people identified as tasty would tell us
about cuisine. Darwin's fundamental insight as a biologist was that
among groups of sexually reproducing organisms, the differences are
much more important than the similarities. If human beings were iden-
tical, a single change in the environment could wipe out the species.
Similarity, ultimately, is death. So why do Darwin's contemporary fol-
lowers want to make what people have most in common into a social
good? The true good is the different, not the same.

That's what Horace Kallen was trying to say, though, back in 1915,
in his article on "Democracy Versus the Melting-Pot"; and as we have
seen, he made that argument by essentializing race and nationality,
making them immutable. Since groups *are* different, since you can't do
away with difference without bringing down the level of the whole,
then we must account difference a social good and preserve it: that is
basically what Kallen said. So if we value difference and deviance on
Darwinian principles, as well as on principles of fairness and tolerance
and even (though it is in short supply today) humility, how do we avoid
doing what Kallen did, and root those differences in biology? Rooting
them in culture is no better: both explanations, biological and cultural,
tend to be deterministic, and determinisms are false not because behavior is not determined, but because it is overdetermined. No single cause
accounts for the whole.

Another way was found by a friend of Horace Kallen's, a fellow
Harvard graduate named Alain Locke. Locke's situation was not exactly
parallel to Kallen's. Locke's situation was not exactly parallel to anyone's. He had heart trouble and an unusually slight physique (he was
five feet tall and weighed ninety-nine pounds); he was homosexual;
and he was black. He had come to Harvard from Philadelphia, where
his parents were schoolteachers, and where he had been a brilliant student in mostly white schools. His undergraduate career at Harvard was
similarly distinguished. But he was careful not to associate too much

with other black students at Harvard because he regarded his life as an experiment in blocking out physical accidents like race. He was the first African American to win a Rhodes Scholarship, and the fact received considerable attention, but it was not how he wished to be known. "I am not a race problem," he wrote to his mother after winning the Rhodes. "I am Alain LeRoy Locke."[6]

When Locke arrived at Oxford on his Rhodes, though, his race did become a problem. Five Oxford colleges denied him admission, and the Southern Rhodes Scholars in his class, who had already formally appealed to the Rhodes trustees to overturn Locke's award, shunned him. Locke found himself the personal focus of racial politics, and he was taken up by nonwhite colonial students from India, Natal, Egypt, and Ceylon. The experience was traumatic, and it gave Locke a much richer appreciation of the social salience of race. He left Oxford without taking a degree. By the time he returned to the United States in 1911 and secured a teaching position at Howard University, he had abandoned the notion that racial difference was a fact of life one could ignore.

In 1915, the same year that Kallen's "Democracy Versus the Melting-Pot" was published, Locke gave a series of lectures at Howard called "Race Contacts and Interracial Relations." He began by citing the work of the man who had helped modify some of the conclusions people had drawn from Weissmann's experiment. This was Franz Boas, one of the fathers of cultural anthropology, and therefore, in a sense, the man who introduced cultural polytheism to the world. Boas had shown, in a physiological study of immigrants, that environment does have biological effects; and Boas was himself very much a cultural pluralist.

Locke argued that there was a distinction between difference and inequality. Racial difference is biological and racial inequality is social, but they are constantly confused. As Boas had said, it is illogical to prevent a group from developing a civilization and then to attribute its failure to develop a civilization to biological inferiority; but that is what Europeans had done to nonwhite races around the world. They had created a history of racial invidiousness, and then they had called it natural. Locke concluded—a conclusion drawn from his own experience at Oxford—that individuals are the bearers of that history, whether they choose to be or not. "When the modern man talks about race," Locke said, "he is not talking about the anthropological or biological idea at all. He is really talking about the historical record of success or failure of an ethnic group. . . . [T]hese groups, from the point of view of anthropology, are ethnic fictions."[7]

They are fictions whose effects are real enough, however. What Locke proposed was a way to make the fiction useful for minority ethnic groups. He did not think that those groups could improve their situation by maintaining separateness, as Kallen had advised in the case of European immigrants. For modern civilization does not tolerate separateness. "Modern systems are systems that require or seem to require social assimilation," as Locke said. People may eat their ethnic food and wear their ethnic hats, but in the things that matter, they are obliged to adhere to the dominant standard. Modern societies, Locke said, "are not necessarily so arbitrary about their social culture as . . . earlier societies were, but they are at least arbitrary to this extent: that in the interests of what they call a common standard of living, common institutions, and a common heritage, they exact that a man who elects . . . to live in a modern society must adopt, more or less wholesale, the fundamental or cardinal principles of that social culture."[8]

Still, if it is a mistake to cling to ethnic identity, it is also a mistake to abandon it. The trick is to use it in order to overcome it. "The group needs . . . to get a right conception of itself," Locke said, "and it can only do that through the stimulation of pride in itself. Pride in itself is race pride, and race pride seems a rather different loyalty from the larger loyalty to the joint or common civilization type. Yet . . . through a doctrine of racial solidarity and culture, you really accelerate and stimulate the alien group to rather more rapid assimilation of the . . . general social culture, than would otherwise be possible."[9] Although racial identity has no basis in biology, and although racial pride is, by itself, socially divisive, the only way to overcome social divisiveness is to stimulate racial pride, to encourage minority ethnic groups to take satisfaction in their particular practices and achievements. The desire to be accepted as like everyone else—the desire to meet the "common standard"—flows from the desire to be recognized as different from everyone else. You want to prove that your group is as good as every other group. The elegance of Locke's formulation is that neither human sameness nor human difference is treated as real and essential. They are defined functionally. Universality and diversity are both effects of social practice. They are not given in nature; they are outcomes of what people do.

Horace Kallen and Alain Locke were both students of William James, who is best known as one of the founders of the philosophy known as pragmatism. Pragmatism today is associated by some people with the denial of truth and objectivity, and it therefore can seem anti-empirical and anti-scientific. But the two most important figures in the development of pragmatism were both trained as scientists—William James and his friend Charles Sanders Peirce. James began teaching at Harvard in 1874 in the physiology department. His field was experimental psychology. He came to philosophy late in his career.

Pragmatism is not a debunking of the concept of truth. It is an effort to adapt the concept of truth to the universe Darwin described, a universe where things happen higgledy-piggledy, where reality doesn't sit still long enough for us to form an accurate picture of it. Pragmatism regards truth in the same way that statistical science regards a fact: it is the provisional place that the preponderance of experience leads us to assert a belief. Truth is like a natural law in science: it is what, given certain conditions, will happen *most of the time*. It is not an iron law. The universe being what it is, there may be a natural law of iron, but there are no iron laws of nature. As one nineteenth-century philosopher of science put it: "Scientific laws are the bed over which passes the torrent of facts; they shape it even as they follow it. . . . They do not precede things, they derive from them, and they can vary, if the things themselves happen to vary." The tendencies of living beings to follow predictable paths "can look, viewed from outside, like necessary laws," but they are only habits. Without variation, everything would be dead matter.[10]

Pragmatists also believed that on a theory of natural selection, there is no warrant for the notion that our minds are supposed to mirror reality objectively. This isn't just because reality doesn't stand still. It's because there is no adaptive utility in having a mirror in our heads. The evolutionary value of having minds is the same as the evolutionary value of having opposable thumbs: it helps us cope with our environment. The truth-value of a belief, therefore, is the same for a pragmatist as the truth-value of a statistical fact is for a scientist: its predictive usefulness. True beliefs, James liked to say, are beliefs that cash out in experience. One of his favorite examples was belief in an idea central to scientific inquiry: causation. You cannot *show* causation, James said, any more than you can show the existence of God. But belief in causation is warranted because experience shows that it pays to believe in causation.

This was a view about truth drawn directly from science. One of James's students, the future educational psychologist Edward Thorndike, had, for his doctoral thesis, put chickens in boxes with doors on them. Then he measured how long it took the chickens to learn how to open the doors and get at the food pellets outside. He observed that although at first many actions were tried, apparently unsystematically, only successful actions performed by chickens who were hungry—only actions that opened the door to food the chickens wanted to eat—were actually learned. Actions that produced no results were simply forgotten by the chickens. He concluded that success is what caused those movements to be imprinted in the brains of the chickens. Belief that pushing this lever with my beak will give me access to food is a belief that cashes out in experience. It is therefore, pragmatically, true. Belief that I have to emit a special cluck before I push the lever could be a vestigial belief, discarded when it becomes clear that without the cluck, the door opens anyway.

The pragmatists thought that philosophers had mistakenly insisted on making a problem of the relation between the mind and the world, an obsession that had given rise to the attempt to answer the question, "How do we know?" The pragmatist response to this question is to point out that nobody has ever made a problem about the relationship between, for example, the *hand* and the world. The function of the hand is to help the organism cope with the environment; in situations in which a hand doesn't work, we try something else, such as a foot, or a fishhook, or an editorial. Nobody worries in these situations about a lack of some preordained "fit"—about whether the physical world was or was not made to be manipulated by hands. They just use a hand where a hand will do.

The pragmatists thought that ideas are the same as hands: instruments for coping. An idea has no greater metaphysical stature than, say, a fork. When your fork proves inadequate to the task of eating soup, it makes little sense to argue about whether there is something inherent in the nature of forks or something inherent in the nature of soup that accounts for the failure. You just reach for a spoon. But philosophers have worried about whether the mind is such that the world can be known by it, and they have produced all sorts of accounts of how the "fit" is supposed to work—how the mental represents the real. The pragmatist point was that "mind" and "reality" are only abstractions from a single, indivisible process. It therefore makes as little sense to talk about a "split" that needs to be overcome between the mind and the world as it does to talk about a "split" between the hand and the environment, or the fork and the soup. The pragmatist would make the same point about biological and psychological explanations of personality and behavior: they are abstractions from a single entity, which is the human being. You can speak of the human organism from the point of view of genetics, and you can speak of it from the point of view of psychology, spirituality, or culture. Anything that helps us get a grip on understanding the phenomenon is useful. The phenomenon itself is the sum of all possible understandings.

James was a Darwinian, but he was not a Darwinist, exactly. What he admired about the theory of natural selection was that it did not attempt to ignore the eccentricities of the world, as taxonomical biologists and creationists had tried to do, but built its theory up from them. "It is one of the fortunate points of the general theory which bears [Darwin's] name," James wrote when he was a young man, "that the more idiosyncrasies are found, the more the probabilities in its favor grow," since idiosyncrasies in nature are evidence of the existence of chance variation.[11]

But James thought that people took the wrong lesson from *On the Origin of Species*. This is the belief that we see today in popularized evolutionary psychology, the belief that evolutionary science can lay a foundation for norms, that natural selection serves as a kind of "bottom-line" arbiter of merit. This makes the logic of evolution the logic of human values: it suggests that we should pursue policies and honor behavior that are consistent with the survival of characteristics understood to be "adaptive," and it justifies, as "natural," certain kinds of coercion. It is therefore a scientific theory for winners: it ratifies every triumphant outcome by explaining it as the result of natural selection. It is a free-market philosophy for organisms.

James believed that scientific inquiry, like any other form of inquiry, is an activity inspired and informed by our tastes, values, and hopes. But this did not, in his view, confer any special authority on the conclusions it reaches. On the contrary: it obligates us to regard those conclusions as provisional and partial, since it was for provisional and partial reasons that we undertook to find them. The mistake is not simply endowing science with an authority it does not merit. It is turning one belief into a trump card over alternative beliefs. It is ruling out the possibility of other ways of considering the case. James believed that the theory of natural selection should be regarded like any other idea—as a hypothesis, good in some situations, not so good in others. It should not be regarded as a basis for values. Natural selection is, after all, a chance process. The bird with the better-adapted beak isn't smarter or nobler than the other birds; it just lucked out. A characteristic that helps an organism survive may be completely undesirable from every other point of view, and survival in one season can mean extinction in the next. The real lesson of *On the Origin of Species* for James—the lesson on which he based his own major work, *The Principles of Psychology*, published in 1890—is that natural selection has produced, in human beings, organisms gifted with the capacity to make choices incompatible with "the survival of the fittest." There *is* intelligence in the universe. It is ours. It was our good luck that, somewhere along the way, we acquired minds. They released us from the prison of biology.

After James changed fields from psychology to philosophy, he began developing his idea of pragmatism. In 1907, he published a book with that title, presenting his philosophy to the world. He dedicated it to the British philosopher John Stuart Mill, who, James said, "my fancy likes to picture as our leader, were he alive today." Mill had died in 1873. A year after his death, one of his last essays was published. It is called "On Nature," and I close by quoting from the ending of it.

"The word 'nature,'" Mill wrote, "has two principal meanings: it either denotes the entire system of things, with the aggregates of all their properties, or it denotes things as they would be, apart from human intervention. In the first of these senses, the doctrine that man ought to follow nature is unmeaning; since man has no power to do anything else than follow nature; all his actions are done through, and in obedience to, some one or many of nature's physical or mental laws. In the other sense of the term, the doctrine that man ought to follow nature, or, in other words, ought to make the spontaneous course of things the model of his voluntary actions, is equally irrational and

immoral. Irrational, because all human action whatever consists in alter-
ing, and all useful action in improving, the spontaneous course of
nature. Immoral, because the course of natural phenomena being replete
with everything which when committed by human beings is most
worthy of abhorrence, any one who endeavored in his actions to imitate
the natural course of things would be universally seen and acknowledged to be the wickedest of men. The scheme of Nature, regarded in
its whole extent, cannot have had, for its sole or even principal object,
the good of human or other sentient beings. What good it brings to them
is mostly the result of their own exertions. Whatsoever, in nature, gives
indication of beneficent design proves this beneficence to be armed
only with limited power; and the duty of man is to cooperate with the
beneficent powers, not by imitating, but by perpetually striving to
amend, the course of nature—and bringing that part of it over which
we can exercise control more nearly into conformity with a high standard of justice and goodness."[12]

These are wise words still.

NOTES

1. *Charles Darwin's Marginalia*, Vol. 1, ed. Mario A. di Gregorio (New
York: Garland, 1990–), 164.

2. Horace Kallen, "Democracy Versus the Melting-Pot," *Nation* 100
(1915): 220.

3. Horace Kallen, *The Structure of Lasting Peace: An Inquiry into the
Motives of War and Peace* (Boston: Marshall Jones, 1918), 31.

4. Kallen, "Democracy Versus the Melting-Pot," 220.

5. Steven Pinker, *The Blank Slate: The Modern Denial of Human Nature*
(New York: Viking, 2002), 50.

6. Alain Locke to Mary Locke, March 23, 1907; quoted in Louis Menand,
The Metaphysical Club (New York: Farrar, Straus and Giroux, 2001), 390.

7. Alain Locke, *Race Contacts and Interracial Relations: Lectures on the
Theory and Practice of Race*, ed. Jeffrey C. Stewart (Washington, D.C.: Howard
University Press, 1992), 12.

8. Locke, *Race Contacts and Interracial Relations*, 91.

9. Locke, *Race Contacts and Interracial Relations*, 96–97.

10. Emile Boutroux, *De la contingence des lois de la nature*, 2nd ed. (Paris:
Ancienne Librarie German Baillière, 1895), 39, 167. My translation.

11. William James, "Two Reviews of *The Variation of Plants and Animals under Domestication*, by Charles Darwin" (1868), *Essays, Comments, and Reviews*, in *The Works of William James*, ed. Frederick Burkhardt (Cambridge, Mass.: Harvard University Press, 1975–88), 234–35.

12. John Stuart Mill, "On Nature" (1874), *Nature, the Utility of Religion and Theism* (London: The Rationalist Press, 1904), 32–33.

THE CULTURAL CONTEXT OF DISABILITY

Nora Groce

Disability is a universal—in all societies, through all ages, individuals have been born with or have acquired disabilities. The lives of these individuals, however, will in large measure be determined not by the fact that they have a disability but by the way the society in which they live conceptualizes what it means to be "disabled." While discussion of genetics and specific physiological, psychological, intellectual, and sensory impairments can be done without reference to "culture," understanding what it means to live with a disability requires an understanding of the cultural context in which a person with a disability exists.

It is important to underscore the fact that disability has a strong cultural component, because this has so often been overlooked. Whole libraries are devoted to the medical, rehabilitation, and psychological implications of particular disabilities on the individual. Discussion in these disciplines is sometimes expanded to include considerations of the psychology, education, or employment of the individual who has the disabling condition. But people who are disabled do not live in a vacuum. All individuals with disabilities live within a specific culture, and they share, with other members of their culture, learned behaviors, beliefs, attitudes, values, and ideals, which characterize their society. All individuals with disabilities are also members of a social network—

they are sons and daughters, fathers and mothers, husbands and wives, cousins, neighbors, and fellow citizens.

Cross-cultural comparisons of disability—how the same type of disability is interpreted differently depending upon the culture in which a person lives—provide interesting insight. To do a cross-cultural analysis of disability is to look at disability as the constant and the cultural context in which it is found as the variable. The cross-cultural study of disability is a relatively new field. Prior to the 1980s, most scholars assumed that there was no cultural variable—that all societies at all times reacted in much the same way to persons with disabilities. Moreover, there was an assumption that if persons with disabilities fared badly in modern, wealthy societies, their lives must be much more difficult in more traditional societies or in countries where modern medicine and rehabilitation are not widely available. Although, as we shall see, there is an enormous variation in how individuals with disabilities are incorporated into the social life of a community, in many cases, traditional and/or nonwestern models of adaptation to disability have much to teach us about universal approaches to disability.

BACKGROUND

It is important to note that we still know relatively little about disability in any society. Although there is a growing literature on disability cross-culturally, the majority of the published research is from developed nations; and most of this work itself is based on studies of urban populations. Yet it is estimated that 80 percent of all individuals with disabilities today live in the developing world and of these, 60 to 70 percent live in rural areas.

One of the most striking findings of this growing body of sociocultural research has been the discovery that around the world, there are few ideas about disability that are held to be true at all times and by all people. In fact, there are considerable differences in the way disability is regarded from one society to the next. Even within the same society, different types of disabilities tend to be regarded differently; an individual who is Deaf may be considered a full and active participant of a community, whereas an individual with mental health problems who lives next door may be shunned.

One conclusion can be drawn from this growing body of data: the lives of individuals with disabilities around the globe are usually far more limited as a result of prevailing cultural constraints and the attendant social, cultural, and economic ramifications of these cultural constraints than as a result of any specific physical, sensory, psychological, or intellectual impairments.

SOCIAL INTERPRETATION OF DISABILITY

Individuals with disabilities have always been part of human society, and no society has been found anywhere in the world that does not have a complex system of beliefs about disabilities. In all societies, it is recognized that there are some individuals with certain attributes that distinguish them from other members of that society. Universally, societies have explanations for why some individuals (and not others) are disabled, how individuals with disabilities are to be treated, and what roles are appropriate (and inappropriate) for such individuals. All societies also have a strong set of beliefs about the rights and responsibilities to which individuals with disabilities are either entitled or denied.

The issue, of course, is not only that individuals with disabilities exist, but also how the societies in which they live conceptualize what this existence should be like. Societies do more than simply recognize disabling conditions in their members; they attach value and meaning to various types of disabilities. Although dozens of different examples exist, it is important to approach this body of data with some coherence. For the purposes of this discussion, these social beliefs are grouped together into the following three categories, which seem to regularly appear cross-culturally:

1. Beliefs about causality: the cultural explanations for why a disability occurs;
2. Valued and devalued attributes: specific physical or intellectual attributes that are valued or devalued in a particular society; and
3. Anticipated role: the role an individual with a disability is expected to play as an adult in a community.

These categories seem to be used consistently across cultures as a basis upon which people's expectations and demands for (or avoidance of and passivity about) how individuals with disabilities are treated. These categories also seem to affect individuals' ability to participate in family as well as a community's and society's willingness to integrate these individuals into daily life. Although there are many variations, the following is an overview of the more salient issues involved in each of the above categorizations.

Beliefs about Causality

Cultural beliefs about why a disability occurs help determine how well or how poorly societies treat these individuals. For example, in some cultures, the birth of a child with a disability is considered a sign of divine displeasure with the child's parents, evidence of incest or "bad blood" in a family, evidence of marital infidelity, or the result of bad luck or fate.

Reincarnation, the belief that one's current physical and social state is a reflection of one's behavior in a previous life, often leaves individuals who are disabled in a particularly difficult situation. Their current status is seen as earned, and therefore there may be less sympathy and less willingness to expend resources on their behalf. Moreover, improving their present lives, some believe, lessens the amount of suffering they must endure, thus compromising the possibility of future rebirth at a higher level of existence.

The cause of disability is not always believed to be divine or supernatural. The idea that a disability can be "caught"—transmitted either by touch or by sight—is found widely. Modern science has redefined disability causation, seeking explanations in the natural world. Genetic disorders, viruses, and accidents are now commonly accepted as explanations for a congenital or an acquired disability. But if modern medicine has replaced older causation concepts, it has often not done so completely. The idea of blame, inherent in most cultures for centuries, often reappears in more "scientific" forms. Thus, for example, in modern Western societies, both professionals and laypeople are quick to question whether the mother of a disabled newborn smoked, consumed alcohol, or took dangerous drugs, even if there is no correlation between such behaviors and the specific disability of her child.

The need to identify a reason for the appearance of a disability, some have speculated, may be a psychological distancing. Individuals try to establish a logical reason why a disability has occurred to someone else, thus reassuring themselves that something similar will not happen to them. Another reason why there is such widespread attention to causality may be society's attempt to determine what demands the individual with a disability, as well as that individual's family, may justifiably make on existing social support networks and community resources.

However, beliefs that link disability to intentional causality are not always negative. A study of parents of children with disabilities from northern Mexico, for example, found that these people believe that it is God's will that a certain number of children with disabilities be born. God, being kind however, chooses parents who will be especially loving and protective to these "special" children.

Valued and Devalued Attributes

The personal attributes a society finds important are perhaps even more important in predicting how well an individual with a disability will fare in a given society than are beliefs about how a disability is caused. Those individuals who are able to manifest or master valued attributes will be able to play a broader role in their societies than those who cannot or who can do so only with difficulty. This will, in turn, be reflected both in the manner in which these individuals are treated and in the society's willingness, or unwillingness, to allocate resources to meet their needs. For example, in societies where physical strength and stamina are valued, where one's status in the community depends in large measure on how well one can fish or farm, difficulty in walking or in lifting will diminish one's social status. Conversely, and increasingly, in societies where intellectual endeavors—the ability to work at an office computer or to research and prepare a speech—are considered important, the fact that one uses a wheelchair will be far less significant.

This variability in valued attributes is strikingly apparent in the case of gender. In societies where boys are preferred, the willingness of families to expend scarce resources on a girl with a disability might be substantially less than for a comparably disabled boy. Issues of social class, economic status, family structure, level of education, and more will also have implications for valued and devalued attributes. These variables will affect what individuals and families hold to be important and will enable some individuals within any given culture to more fully develop their talents and abilities. For example, in a society where education is held to be very important, a child with a disability from a wealthy family may be at a significantly greater advantage than a child with a comparable disability from a poor family. Better schools, the ability to hire tutors, and the means to purchase assistive equipment such as a computer or more attendant care may enable the wealthier disabled child to excel.

Anticipated Role

Finally, the willingness of any society to integrate individuals with disabilities into the surrounding culture, including the willingness to expend resources for education, health care, skills training for jobs, and so forth, will also depend in large measure on the role or roles individuals with disabilities are anticipated to play in the community as adults.

At one extreme, a society might anticipate no adult roles and refuse to allocate any resources for those with disabilities. In such societies, theoretically, children born with a disability might not be allowed to live. Infanticide of even severely disabled newborns, however, is exceptionally rare in the ethnographic literature. In more recent years, the use of amniocentesis, genetic counseling, and the withholding of medical care in the delivery room, while touted as medical advances by some, are viewed by many as more technologically sophisticated (and more widely practiced) forms of infanticide.

Survival is not the only measure of cultural inclusion. In some societies, individuals with disabilities are kept alive but hardly welcomed. Some have argued that in many societies, an individual's inability to have a paid job, and thus to contribute to his or her family's economic well-being, is the deciding factor as to what status he or she maintains in the household and in the community. Calculating a person's economic contribution to his or her family or society in terms of formal employment, even marginal employment outside the home, may be misleading, however, and such data must be used with caution. Many individuals with a disability who do not work outside their own homes or family units make significant contributions to their family's economic well-being. All but the most significantly disabled individuals often make contributions. They watch children, cook, clean, do housework and farm work; they help assemble piecework or do crafts that are brought to the marketplace in someone else's name.

Individuals with disabilities are often allowed only partial inclusion into society and given limited roles and responsibilities. Traditional professions for individuals with disabilities are often reported. For example, in some societies, blind people become musicians, potters, or broom makers, and mobility-impaired individuals work as market vendors. In other societies, individuals with disabilities are thought to be inspirational and, although ill-treated on a day-to-day basis, at certain times of the year or on certain ceremonial occasions, become the center of attention. Christmas is an example of such an occasion in the West.

A full adult role in any community implies not only employment but also the ability to marry, have a family of one's own, and decide where one will live, with whom one will associate, and how one will participate in the civic, religious, and recreational life of the community. Although societies differ as to where, when, and how individuals carry out these roles, the issue is whether individuals with disabilities are participating in such activities at a rate comparable to that of their nondisabled peers. Full acceptance, that is, status and treatment comparable to one's nondisabled peers, is relatively rare, but it does exist and is important.

Indeed, communities may interpret even significantly disabling conditions in a positive light. For example, on the island of Martha's Vineyard, off the northeast coast of the United States, a gene for profound hereditary deafness led to the birth of a number of Deaf individuals from the mid-seventeenth to the late-nineteenth centuries. Because deafness was so common, it was in the best interests of the hearing islanders to learn and use sign language, and most did. With the breaching of the substantial communication barrier—the very thing that most regularly blocks Deaf individuals from full participation in society—it is perhaps not surprising that Deaf individuals on Martha's Vineyard participated vigorously in the life of the small villages in which they lived. They were not considered to be (nor did they consider themselves to be) disabled. The fact that individuals with disabilities assume roles comparable with all other members of a society is a good indication that real integration has been achieved.

MODERNIZATION

The social role that an individual holds in society may well change over time as societies modernize. New technologies will provide workplace options to many individuals with disabilities, but additional barriers to some. For example, wheelchair users have an increasing number of choices in an urban area where transportation, buildings, and sidewalks are increasingly accessible. The growing number of jobs, even very low skills jobs that require literacy and the ability to use computers and other complex machines, however, may limit the employment options of those with some types of intellectual impairments. Indeed, their ability to compete in the workforce as adults may be considered so compromised that in many countries, they are now placed on a formal pension system at the age of eighteen—a system that will maintain them for the remainder of their lives.

CONCLUSION

The contention of this chapter is that while physical, intellectual, sensory, or mental health impairments are universal, the experience of being disabled is largely shaped by the culture in which one lives. The examples provided are far from exhaustive, but it is hoped that an awareness of this cross-cultural variation in approaches to disability will interest readers enough so that they begin to investigate—and continue to watch—this rapidly expanding area of research.

PART 2

DEAFNESS AND GENETICS: A TROUBLED PAST

INTRODUCTION

Historically, the link between genetics and deaf people was provided by eugenics, by the attempt to "improve" society through the elimination of particular genes believed to be deleterious to social progress. American and English authors were already speculating about the possible genetic basis of deafness by the mid-nineteenth century. They wondered whether consanguineous marriages might explain genetic deafness and should, therefore, be discouraged or prohibited. Nearly all historically influential discussions of eugenics and deafness, however, begin with Alexander Graham Bell, who popularized the link between genetics and deafness, argued that deaf people represented a cultural and economic drain on society, and advocated various measures to reduce the incidence of genetic deafness. Similarly, historical perspectives on eugenics and deafness, or disabilities of any kind, must include an examination of Nazi

31

practices during the 1930s and early 1940s, when, as John Schuchman writes in one of the essays in this section: "Medical doctors, lawyers, and educators of deaf children assisted in subjecting deaf Germans to sterilization, abortion, marriage restrictions, and sometimes murder."

The relationship between the American Bell, the great popularizer of fears about the creation of a "deaf variety of the human race," and the devastation of the German deaf population under National Socialism's eugenics program is complex and historically difficult to assess. Two essays in this section look at this issue another way, that is, they suggest why eugenics did not have a greater impact on the American deaf community.

In the first, deaf historian Brian Greenwald offers a revisionist interpretation of Bell. He reviews Bell's role and influence within the American eugenics movement and shows that Bell had the respect of the most prominent American eugenicists. Bell's success as the inventor of the telephone and his research on both technology and genetics earned him acceptance within the American scientific community. His intimate knowledge of deafness, from personal experience with his mother and wife and from his studies of deaf people on Martha's Vineyard, caused American eugenicists to defer to him on matters related to the deaf population. Greenwald argues, therefore, that Bell could have been extremely destructive to deaf people's right to marry and reproduce as they wished. The opportunity was available for Bell to advocate invasive government eugenic measures against the American deaf population, but he did not do so. Greenwald believes that several factors explain Bell's behavior, but he concludes that Bell's personal contact with deaf people throughout his life "humanized and personalized" his approach.

Joseph Murray agrees in his article about the debate over deaf-deaf marriages and writes about Bell's complex struggle to come to terms with his deaf student George Sanders's choice of a genetically deaf woman for a wife, but the main emphasis of Murray's essay is the role of deaf people themselves in countering eugenic activities. He looks at deaf people in the United States, England, and northern Europe (primarily France) as comprising a transatlantic community that worked together on common interests. Among these was maintaining the right to marry whom they pleased. Murray is especially interested in the arguments that deaf people used to defend marriage rights, and he shows that they not only used "scientific" data to support their cause but also appropriated common social beliefs of the time, as well. Middle-class deaf men emphasized the importance of "male autonomy in private affairs," individual choice, and the significance of "true love" for a successful marriage. The result, Murray writes, is that, despite the eugenic opposition to deaf intermarriage in the early twentieth century, deaf people maintained their "traditional marriage rights."

In Germany, however, eugenic considerations took away more than deaf people's right to marry. Under National Socialism, which counted many deaf proponents, deaf people were considered to be a threat to the state, and their elimination became public policy. Deaf children were forcefully taken from school and sterilized; deaf women had their fetuses aborted; eventually, some deaf people were killed. Schuchman's essay concludes this historical section by showing what did happen and, therefore, what can happen. Schuchman concludes that "eugenicists could not anticipate the abuse that occurred," but they created the "milieu of disrespect for disabled people in general, including the deaf population," which provided the context for Nazi practices.

THE REAL "TOLL" OF A. G. BELL: LESSONS ABOUT EUGENICS

Brian H. Greenwald

The relationship between Alexander Graham Bell and other eugenicists during the late-nineteenth and early-twentieth centuries is worth studying to understand the intertwining of deafness and eugenics in early-twentieth-century America. I will approach this subject by reviewing Bell's involvement with eugenics and then examining the ramifications of this historical episode for the Deaf community.

Many people are familiar with Bell's advocacy of oral education for Deaf people, as well as the challenges that his actions posed to our cultural community, but some scholars exaggerate Bell's role in the discussion of marriage among people who were thought to have hereditary deafness. In so doing, they fail to understand clearly Bell's position in the eugenics movement. In addition, they often overlook the historical times in which Bell lived and worked.

During the Progressive Era (1900–1917), the eugenic gospel reverberated across the United States. Indiana passed the first sterilization law in 1907, and other states followed. Federal immigration laws, such as the Chinese Exclusion Acts of 1882 and 1902, the Gentleman's Agreement of 1907, and the National Origins Act of 1924, showed eugenic influences. Eugenic ideals filtered down to local and regional areas, too: the Galton Society met periodically at the American Museum of Natural History in New York, and John H. Kellogg established the Race Betterment Foundation in Battle Creek, Michigan. Eugenic Education Societies proliferated in states such as California, Minnesota, and Wisconsin, and in cities such as Chicago and St. Louis. Influential Americans such as Charles Benedict Davenport; David Starr Jordan, president of Stanford University and chairman of the Eugenics Section of the American Breeders Association (ABA); Luthur Burbank, a well-known horticulturist; Henry Fairfield Osborn, president of the American Museum of Natural History; and Bell unabashedly promoted eugenics. Organizations and universities served as breeding grounds for the next generation of eugenicists.

It is in this context that we see Bell as very much a man of his times. Bell was a Progressive reformer who advocated women's suffrage. He was technologically savvy, as his many inventions demonstrate. Like some Progressives, Bell held strong anti-immigrant attitudes, even though he emigrated from Scotland to the United States and became a naturalized citizen himself.

Bell espoused "positive" eugenics regarding Deaf people. Positive eugenicists encouraged procreation among those who were considered "good in stock," or genetically "fit." By contrast, negative eugenicists sought to stop the spread of "bad genes" through invasive measures, such as mandatory placement in institutions, sterilization, or the prohibition of marriages between those considered "weak in stock" and genetically "unfit." The misguided belief that Bell advocated the sterilization of Deaf people has generated outrage and distracted people from a far greater threat to the Deaf community, that is, his staunch support of oralism.

In 1883, Bell delivered an address to the National Academy of Sciences entitled *Memoir Upon the Formation of a Deaf Variety of the Human Race* (henceforth referred to as the *Memoir*). Some perceive this as a ruthless attack on deaf-deaf marriages; yet others interpret this essay as Bell's recognition of the *right* of Deaf people to marry. This polemical document identified factors that caused the growth of Deaf culture, including the isolation of Deaf people, the institutionalization of Deaf children, sign language, and intermarriage between Deaf adults.

In the *Memoir*, Bell noted that one option to stem the growth of Deaf culture—which was Bell's goal—was legislation prohibiting the intermarriage of congenitally deaf people. A second option was the utilization of "preventive measures," such as the elimination of deaf schools, removing sign language from the curriculum, and eliminating deaf teachers and administrators.

Bell did not support the outright prohibition of intermarriage among Deaf people. Prohibition would not cease sexual relations among deaf men and women, he believed. The result of a law against Deaf intermarriage, therefore, would be that illegitimate children would abound. As a man of principles, Bell's moral beliefs could not accept this alternative. Instead, he encouraged Deaf people to resist marriage to each other, and he proposed deaf-hearing marriages as an alternative. Bell thought that such unions would cause the number of deaf offspring to decline. Therefore, although Bell recognized the right of deaf people to marry, he thought deaf-hearing marriages would achieve his goal of stripping Deaf people of their cultural identity and their tendency to intermarry and produce more Deaf children.

After Bell delivered his 1883 address, he explored other channels of discourse and thinking on eugenics. Bell came to the Gallaudet College campus in 1891 to deliver an address to the Literary Society. He respected Deaf people as capable thinkers. In that address, Bell attempted to dismiss the myth that he wanted to prevent deaf people from marriage, stating that he had no desire to "interfer[e] with your liberty of marriage . . . [because] I myself, the son of a deaf mother, have married a deaf wife."[1] From that point on, Bell did not return to address the Deaf community personally, except for his "few thoughts concerning eugenics" in 1908.[2] Bell's eugenic study as well as statements about deafness continued long after the *Memoir* and the Literary Society speech, however. These provide a more complex image of Bell and his view of our community.

Not long after the 1891 address to the Literary Society, Bell became much influenced by the works of Charles Darwin. On his Beinn Bhreagh estate in Nova Scotia, Bell kept numerous pairs of rams and ewes to determine if multinippled sheep were more fertile than normal two-nippled sheep. He worked to determine whether the sheep with extra nipples produced a greater quantity of milk. Bell killed all of his four-nippled rams, keeping only the six-nippled rams for breeding.[3]

Bell was thorough and meticulous in his data, identifying each sheep by number, recording sheep weight in the fall and winter months, making nipple measurements, and he even tracked deformed sheep. Bell continued his sheep experiments until his death, for he loved to experiment. This interest in genetics and breeding brought Bell into close contact with other eugenicists sharing similar interests.

The most important eugenicist was Bell's close friend Charles Benedict Davenport, the founder and director of the Station of Experimental Evolution at Cold Spring Harbor, New York. Davenport was interested in Bell's sheep experiments and even purchased some of Bell's ewes and shipped them to Cold Spring Harbor for further study. Thus Bell's friendship with Davenport blossomed and brought him into the broader eugenics movement during its heyday, 1895 to 1930.[4] The timing was auspicious.

Davenport, the most prominent eugenicist in the United States during the late-nineteenth and early-twentieth centuries, recognized Bell as the nation's leading scientific expert on genetic deafness on the basis of his previous work in the *Memoir* as well as his studies of hereditary deafness on Martha's Vineyard during the 1890s. Furthermore, Davenport was fully aware of Bell's research skills and trusted that Bell could help solve the "problem" of hereditary deafness. The authority bestowed on Bell by Davenport and the other prominent eugenicists of that period had critical implications for the fate of the Deaf community.

In December of 1908, therefore, Davenport wrote Bell that he wanted to study deafness. If records could be maintained, Davenport reasoned, determining the "proportion of offspring of a particular mating of deaf strains which will be particularly liable to deafness" would enable eugenicists to "discover a precise law of mating ensuring normal offspring from a parent with hereditary tendency towards ear defect."[5] Davenport then offered Bell the chairmanship of the subcommittee of deafness of the Committee of Eugenics under the governance of the American Breeders Association (ABA). The first organization promoting eugenics, the ABA was founded in 1903, and it charged the Committee of Eugenics to "investigate and report on heredity in the human race" in order to determine "the value of superior blood and the menace to society of inferior blood."[6] Bell accepted Davenport's offer.

Throughout Bell's association with the ABA, he speculated on various plans to improve human breeding, and he continued to support eugenics. In 1912, for example, Bell praised his friend Davenport, writing, "You have started a great work, of vast importance to the people of the United States and to the world, by the establishment of the Eugenics Record Office, and I can assure you of my hearty co-operation as one of the Board of Scientific Directors."[7] During the 1911 ABA meeting, and at Bell's invitation, the Committee of Eugenics met at the Volta Bureau, which Bell had established in Washington, D.C., to promote the oral education of deaf children. Many renowned eugenicists converged at oralism's landmark institution. There, Bell talked about the foundation and work of the Volta Bureau.

Bell offered oralism as an attractive option to sterilization and marriage bans for promoting the eugenic goal of reducing the incidence of hereditary deafness. The *Memoir* posits oralism as the instrument for Deaf success and the route to a mainstream life for deaf individuals. The confluence of his beliefs and influence—pedagogy (oralism) and science (eugenics)—allowed him to shape public policy for the Deaf community. In this way, Bell asserted his "authority" on matters related to deaf people. Practitioners of negative eugenics understood that deafness was Bell's domain and did not interfere with his work.

Bell's continuing importance was reflected by his selection as the honorary president of the Second International Congress of Eugenics in 1921, only one year before his death. Davenport encouraged Bell to exhibit his work on the longevity of offspring or on the deaf families of Martha's Vineyard at the Second Congress.[8] In response, Bell sent a display of "six stereograms showing the relation between Age of Fathers at Death, Age of Mothers at Death, and Longevity of Offspring."[9] Davenport, Henry Fairfield Osborn, Clark Wissler, and Harry H. Laughlin later published the Congress proceedings and recognized Bell as a "pioneer investigator in the field of human heredity."[10] Until his death in 1922, Bell met with eugenicists at his Washington, D.C., home and utilized the Carnegie Institution as well as the Volta Bureau as meeting sites when Davenport, Kellogg, and other eugenicists traveled to the nation's capital.

In summary, Bell was an influential and respected figure within the American eugenics movement. Eugenicists such as Davenport, Charles Eliott, and David Starr Jordan recognized Bell as the "expert" on deafness due to his professional work, his personal experiences with a deaf wife and mother, and his experience as a teacher of deaf children. He could have encouraged the development of harshly negative eugenic steps against the American Deaf community. At the very least, he could have stepped aside, allowing eugenicists to pursue their interest in the genetics of deafness. Certainly this would have had far more damaging implications for the Deaf community than Bell's support of oralism. I suggest, therefore, that Bell kept his friends in scientific circles—nearly all of whom could be labeled as negative eugenicists—in close check. It's worth speculating briefly on why Bell was not eager to apply negative eugenic remedies to Deaf people.

First, it should be understood that negative eugenicists targeted feebleminded individuals, singling them out for sterilization or permanent placement in institutions. They argued that allowing these individuals to procreate would undermine the strength of American society. Bell knew that Deaf people were not feebleminded, and he did not confuse lack of intelligible speech or poor English language skills with mental retardation. During the heyday of the eugenics movement, being "normal" was paramount. Oralism existed before *eugenic thought* became popular, though, and it was attractive because it promised to make deaf people "normal." Bell's deaf wife practiced oralism, and he considered her "normal" because of that. Bell could conclude, then, that marriage bans and the sterilization of deaf people were unnecessary. "Normality" could be achieved in other ways.

Second, Bell was a man with principles, and he believed that the government should not interfere with Deaf people's "pursuit of happiness in marriage."

Finally, Bell, who could both sign and fingerspell, had family and friends who were deaf. This contact with Deaf people humanized and personalized his view of deafness, distancing him from negative eugenicists on this subject.

What can we learn about the history—and the future—of the Deaf community from this episode?

Deaf leaders in the late-nineteenth and early-twentieth centuries did not appreciate the danger they faced from eugenicists. They failed to comprehend how oralism fed on eugenic ideas and how it reflected popular social ideals during the late-nineteenth and early-twentieth centuries, when mainstream values merged eugenics with rising nativist sentiments. The Deaf community proved it could defeat strong oralist stances by protecting and promoting its culture. However, the scientific corner—eugenics—was much more difficult. Alexander Graham Bell gave us a close example of how potent these combined forces of education and research are. The idea that Bell chose not to attack our community more forcefully and intimately was to the Deaf community's benefit. Bell, although vilified for his *Memoir*, may have actually shielded Deaf people from negative eugenicists. Whether we accept this argument is for future consideration; however, we must acknowledge the realities of the field of genetics and its achievements in the new millennium. Scientists of the twenty-first century, however, may not appreciate the human side of Deafness. As a community, we must learn from the successes and limits of our previous strategies.

NOTES

1. Alexander Graham Bell, "Marriage: Address to the Deaf." Volta Bureau (Washington, D.C.: Gibson Bros, Printers and Bookbinders, 1891), 3–4.

2. Alexander Graham Bell, "A Few Thoughts Concerning Eugenics," *The National Geographic Magazine* (February 1908): 119.

3. Bell to Davenport, 11 March 1904. American Philosophical Society, Charles B. Davenport Papers, B/D27. Henceforth CBD Papers.

4. Bell passed away in 1922. The eugenics movement in the United States declined after 1935 due to horrific medical experiments in Nazi Germany.

5. Davenport to Bell, 8 December 1909. CBD Papers.

6. Carl N. Degler, *In Search of Human Nature: The Decline and Revival of Social Darwinism in American Social Thought* (New York: Oxford University Press, 1991), 43.

7. Bell to Davenport, 27 December 1912. CBD Papers.

8. Alexander Graham Bell, "The Deaf-Mutes of Martha's Vineyard," *American Annals of the Deaf* 31 (3): 282–84. For additional information on deafness on Martha's Vineyard, see Nora Ellen Groce, *Everyone Here Spoke Sign Language: Hereditary Deafness on Martha's Vineyard* (Cambridge, Mass.: Harvard University Press, 1985).

9. Alexander Graham Bell, *Eugenics, Genetics and the Family: Scientific Papers of the Second International Congress of Eugenics,* Vol. 1, ed. Charles Benedict Davenport (Baltimore: Williams & Wilkins, 1923), Plate 4.

10. Ibid., 2. Like Davenport, Osborn, Wissler, and Laughlin were all committed eugenicists. The four scientists compiled the proceedings of the Second International Congress of Eugenics.

"TRUE LOVE AND SYMPATHY"

The Deaf-Deaf Marriages Debate in Transatlantic Perspective

Joseph J. Murray

"I desire to draw attention to the fact that in this country *deaf-mutes marry deaf-mutes*," said Alexander Graham Bell in his opening presentation to the November 1883 session of the American National Academy of Sciences.[1] The enormity of this fact "consumed the entire morning session" of this mid-year meeting of America's most eminent scientists and thinkers, who were apparently fascinated by the poten-

This essay has benefited considerably from the input of a number of scholars. Douglas Baynton and Linda Kerber gave nuanced, detailed, and always-encouraging feedback on this project from its earliest stages. Johanna Schoen offered in-depth comments on the organization of this essay and shared her knowledge of the U.S. eugenics movement. Jeffrey Cox, Lisa Heineman, Paul Greenough, and Shelton Stromquist offered insightful observations on various drafts. An earlier version of this paper benefited from comments received at the German Historical Institute's 2003 Young Scholar's Forum. I am grateful to this community of scholars for their willingness to share their time and ideas.

tial implications of these marriages "in the formation of a deaf variety of the human race."[2]

Bell's paper provoked an intense decades-long debate within Western Deaf communities and among professionals who worked with Deaf people.[3] British and American educators had long corresponded with one another on Deaf-Deaf marriages; a leading British educator declared in 1857: "it is . . . highly inexpedient that the deaf and dumb should marry with each other."[4] Within the field of Deaf education, Deaf-Deaf marriages was an occasional topic of inquiry, but it was not until Bell's 1883 paper that hereditary deafness came into wider public attention.[5] Bell updated the National Academy at its next three annual meetings, and his ideas found resonance among these scientists and the general public, influenced by growing concern over the purity of national peoples.[6] In the context of heredity science and its practical application in the field of public health, Deaf-Deaf marriages were reconceptualized as a menace to Western society.

While the debate over Deaf-Deaf marriages took place in specific nation-states, the ideas informing these national debates circulated across national boundaries. Belying conceptions of minorities as locked into specific localities and employing purely locally based resistance, Deaf people in the United States and Great Britain also shared ideas with one another to counter opposition to their right to marry whom they chose. Their actions illustrate both the ability of and limits facing a small, geographically scattered minority to retain control of their own lives in the face of larger stigmatizing beliefs. Deaf people in the United States and Britain debated the marriages issue at international conferences, in Deaf community periodicals, and during transatlantic journeys. This exchange of ideas forged a common strategy of response by Deaf people in the United States and Great Britain against attempts to paint them as hereditarily tainted.[7]

Deaf people rejected the notion that a "deaf-mute race" would result from Deaf-Deaf marriages. Their lived experiences suggested that the overwhelming majority of Deaf children came from hearing parents, and a similarly large majority of Deaf couples had hearing children. Indeed, research initiated as a result of the debate would show that more than 90 percent of Deaf couples had hearing children.[8] Deaf people also harnessed tropes of domesticity to present Deaf-Deaf marriages as ordinary middle-class marriages. Claiming that Deaf-Deaf marriages were happier than Deaf-hearing marriages, Deaf people stressed a traditional right of citizenship: the right of men to enter a domestic sphere of their own making.[9] Both of these arguments lent support to a fundamental premise underlying the Deaf response; the perceived threat of Deaf-Deaf marriages did not warrant trampling the individual rights of Deaf people in favor of an alleged national good.

Eugenics was not a clearly delineated scientific field at the time of the Deaf-Deaf marriages debate. Indeed Francis Galton only coined the word *eugenics* in the same year as Bell's New Haven address.[10] What existed in the 1880s and 1890s was a loose international network of corresponding scholars and scientists interested in the transmission of hereditary traits and its impact on particular nations or groups of people. Scientists working on questions of heredity often corresponded with extra-national counterparts, the British-American nexus being particularly noteworthy.[11] Bell was a part of this network. He was interested in questions of heredity and involved in the American eugenics movement from its earliest days. In 1907, Bell was offered the chair of the American Breeder's Association's newly formed Committee on Eugenics. He declined, but agreed to serve on the committee and later chaired a subcommittee on hereditary deafness.[12]

Bells ideas on Deaf-Deaf marriages fit the temper of his times well. In the late nineteenth century, a discourse was emerging that saw traditional liberal beliefs in individual rights and autonomy supplemented by a vision of a collective national body regulated by professional scientists and health workers.[13] The idea of physical degeneration of national societies via the unchecked reproduction of hereditarily inferior classes emerged among reformers in these decades.[14] Reformers suggested various strategies to limit marriage and reproduction among "degenerate" groups of people, as well as foster childbearing among socially acceptable groups. Some advocated restrictions on individual reproductive rights for the good of the nation.[15] Progressive groups, such as women's organizations in the United States and Britain, increasingly utilized the ideas of hereditarianism and eugenic thought.[16]

The Deaf-Deaf marriage debate is an early instance of how scientists in the emerging field of eugenics could lend support to members of an existing social institution—the field of Deaf education—in an attempt to modify the reproductive practices of a particular group. Laws restricting the marriages of those deemed a threat to the body politic were first enacted in the United States in 1896, and they pro-liferated across a number of states over the next decade.[17] The Deaf-Deaf marriages debate took place well before any law was enacted and before any other social group was targeted with specific reproductive restrictions for reasons of heredity. As a small, easily identifiable population, Deaf people seemed the ideal group by which to introduce ideas of hereditary disease and population control into larger public practice. Not only was the general public familiar with deafness, Deaf people were also the indirect recipient of public funds, via state schools for Deaf people. A clear case could be made that a reduction in the number of Deaf people would both save money and eliminate a disability from society. Promoted by Bell's energetic warnings of a "deaf-mute race," the marriages of Deaf people became an object of attention among scientists, educators, and the general public.

To stem the emergence of such a race, Bell proposed a number of measures to inhibit Deaf people from their traditional practice of marrying one another. Bell rejected a legal ban on marriages of Deaf people as impractical, since it would only promote "immorality" among people already associating with one another.[18] The root of the problem of "intermarriages" was "the preference that adult deaf-mutes exhibit for the company of deaf-mutes rather than hearing persons."[19] This preference could be remedied by reducing contact between Deaf children. To compel Deaf children to interact more with hearing children than with one another, Bell suggested the expansion of day schools to replace residential schools for Deaf children. Bell felt sign language promoted the "segregation" of Deaf people from larger society and stressed the education of Deaf children should "entirely discard the use of sign language, and cultivate the use of the vocal organs, and the reading of the lips."[20] From the start, Bell tied the prevention of Deaf-Deaf marriages into his larger agenda, promoted the oral method in Deaf education, and linked both to the health of the nation.

Bell's use of the terms *intermarriage* and *deaf-mute race* to describe the marriages of middle-class white Deaf people latched the idea of degeneracy to Deaf people, evoking stronger images than prevailing notions of deafness as an infirmity were able to do. The generally understood meaning of the word *intermarriage* in this time meant marriages between Europeans and non-European natives of colonial societies or between different castes of non-European populations. By using the

term *intermarriage*, Bell referred to both the interbreeding of an inferior subset of the population and to marriages between members of an
enclosed community, fortifying his conclusions that the Deaf community needed to be dispersed before the specter of a deaf-mute race could
be dispelled.[21] Gathered into a community and intermarrying, Deaf
people were a visible and growing blight on the American national
body.

Deaf Americans, by contrast, saw themselves steadily improving
their status in society. In 1800, no provisions had been made in the
United States for the education of Deaf people, and only a few small,
scattered communities of sign language using Deaf people existed in the
United States.[22] From the founding of the first school for Deaf children
in 1817, educational institutions where sign language was used spread
across antebellum America, and new bilingual Deaf communities formed
in their wake. Local, regional, and national organizations of Deaf people proliferated, and over the decades, American Deaf people developed a community infrastructure to rival, if not surpass, that of their
European Deaf counterparts. America was the home of the National
Deaf-Mute College in Washington D.C., the world's only institution
of higher education for Deaf people. Its graduates were frequently
praised (not the least by Deaf people themselves) as examples of prosperous, intelligent, middle-class Americans, the epitome of the potential restorative power of America's progressive institutions.

Yet, despite great progress, Deaf people across the United States
and Europe faced a number of threats in the closing decades of the
nineteenth century. Signed languages were rapidly losing ground in
educational institutions across the United States and Europe in favor
of a "pure oral" method of instruction, a method emphasizing speech
training as the sole means of teaching the national language. Proponents
of the oralist method, among whom Bell was a leading figure, sought
to eliminate the use of sign language in schools for Deaf people. The
number of Deaf instructors was slowly being reduced at schools for
Deaf people, jobs that had traditionally been held by their best and
brightest. Political activity by Deaf people intent on maintaining sign
language-based education in schools was derided by oralists as lay med-
dling in work better left to professionals. The assumed right of Deaf people to participate in professional discourse in education, their right to
associate with one another for mutual aid and social pleasure, and even

their right to choose their own marriage partners were now being called into question.

What was under threat in the Deaf-Deaf marriage debate was the status of Deaf people, especially Deaf men, as autonomous individuals in public life. Bell specified that he was opposed to legislative restrictions on Deaf-Deaf marriages, but he brought his view on marriages to the attention of governmental bodies on both sides of the Atlantic. For his 1889 testimony to the British Royal Commission on the Education of the Deaf, Bell enlisted some of the most prominent men in American science to write statements in support of his theory, including Simon Newcomb and Edward D. Cope, editor of the *American Naturalist*.[23] Bell also sent copies of his research to members of the U.S. Congress and did little to dampen discussion of legislative restrictions on Deaf-Deaf marriages among scientists, eugenicists, and the general public.[24] Articles in the popular press uncritically reviewed Bell's ideas and declared that the possibility of Deaf-Deaf marriages leading to a deaf-mute race was "a startling possibility in evolution backward in an enlightened age."[25] While others sought to marginalize them according to a particular vision of society, Deaf people portrayed themselves as leading ordinary middle-class lives. The Deaf response in the United States and Great Britain was presented largely from a narrow sector of the Deaf community: well-educated, middle-class Deaf men. These men evoked traditional middle-class beliefs of male prerogative in the domestic sphere, of a man's right to run his family life as he so chose. In the marriages debate, a tradition of male autonomy in private affairs came squarely into conflict with an emerging consensus in scientific circles that control of procreation was essential to ensure a healthy public body.

Deaf people in the United States and Great Britain shared responses that, while originating in specific localities, were nevertheless familiar to Deaf people in other nations and other localities. The 1889 International Congress of the Deaf in Paris dedicated a full evening to the topic of marriage under the heading, "The Deaf in the Family- Marriage- Children." In his signed address to the Paris Congress, Deaf Briton Robert Armour concluded his summary of British objections to restrictions on Deaf-Deaf marriages by telling his audience of Deaf men from nearly a dozen countries, "It is scarcely necessary for me to enlarge on any of these [reasons for opposition] since I know that your own sentiments coincide with mine."[26] Via publications and correspondence, British men knew the arguments of American and European men, and this common transnational knowledge shaped the contours of national and local debates on the marriage question.[27] Moreover, Deaf men were responding to a science that drew upon a common global gird of scientific knowledge; a science acting upon a population sharing a physical characteristic—deafness—which appeared indiscriminately over national boundaries. Thus, Deaf responses were conditioned by similarities in the ideas to which they were responding, as well as similarities in their lives irrespective of nationality.

This sameness lent itself well to the transference of ideas and images from one national setting to another. At the 1889 Paris Congress, Deaf Frenchman Laurent Clerc was exalted by one speaker as "the first deaf-mute to marry a deaf girl."[28] While this speaker's claim was incorrect by several centuries, his use of Clerc is an illustration of the tactical mobility of ideas in the Deaf-Deaf marriage debate.[29] Clerc, who died a decade before the 1889 Paris Congress, had co-founded Deaf education in the United States with Thomas Hopkins Gallaudet, the father of Edward Miner Gallaudet, a contemporary skeptic of Deaf-Deaf marriages. Long prominent in Deaf American and Deaf French hagiographies, Clerc seemed to now have achieved iconic status in the transnational Deaf sphere, his image immediately transferable to other national contexts in the marriage debate. If Clerc, pioneer of Deaf education in the New World, could choose a Deaf spouse, then so could Deaf men everywhere.[30]

The imagery of Clerc was only one of a battery of similar arguments and images Deaf people harnessed in their multifaceted response. While the debate ranged over different social and institutional landscapes in the United States and Britain, the Deaf response in both countries was grounded in three basic arguments. Deaf men countered Bell's and others' statistics with their own data, drawing on their own lives to meet scientific objectivity with their own objective facts. They also portrayed their choice of marriages with Deaf women as a defense of traditional marriage rights, including the rights of Deaf men and women to interact in a private sphere of their own making. Last, Deaf people and their supporters defended a liberal individualist idea of personal autonomy against suggestions that individual happiness ought to be submerged for the good of the nation-state.[31]

Deaf people answered Bell's assertion of a growing "deaf-mute race" by providing contrary statistical evidence. While Bell was able to marshal statements of support from eminent scientists and brandish statistics compiled from years of work, Deaf people provided testimonies from their own lives, setting forth experientially based counterresponses to scientific discourses. A year after the Paris Congress, one Deaf Briton entertained the first National Deaf and Dumb Conference in London with a personal argument for Deaf-Deaf marriage:

> I have married a deaf-mute wife who had also a deaf-mute sister; but though there have been six children born, they are all perfect in their facilities . . . now I have six unanswerable reasons for intermarriage amongst the deaf and dumb.[32]

Deaf people collected qualitative counterevidence showing that Deaf-Deaf marriages only infrequently led to Deaf offspring. Deaf Briton George Healey reported 400 deaf and dumb adults in Liverpool, of whom 118 married one another. "They have 234 children and 60 grand children [sic] and I am thankful to say they can *all* hear and speak perfectly."[33] Statistics showing "most convincing proof of the fallacy of professor Bell's theory" were collected by a number of local communities, with equally skeptical educators and missioners often joining Deaf people in questioning Bell's evidence.[34]

To resolve the issue of competing statistics, American educator Edward Allen Fay undertook a monumental research project to trace the marriages of all Deaf people in the United States. The success of this project depended on the cooperation of Deaf community networks in distributing Fay's surveys, as well as on the willingness of Deaf people themselves to reveal personal details of their lives. On both counts, Deaf people responded enthusiastically, certain that quantitative evidence would validate what they already knew from their own observations. Deaf periodicals were filled with requests to send in information, and one newspaper, the *Deaf-Mute's Journal*, printed detailed lists of Deaf couples.[35] In all, Deaf Americans sent over 4,000 surveys to Fay.[36] The result of this outpouring of information was dissected by Fay, working with May Martin, a Deaf student at the National Deaf-Mute College, and published by Bell's Volta Bureau in 1893.[37] Fay's study validated the Deaf argument, showing over 90 percent of Deaf-Deaf marriages did not produce Deaf children.[38] Fay also noted, ten years after Bell first presented his paper to the National Academy of Sciences, that the overwhelming majority of Deaf people married other Deaf people.[39] More than a decade after Bell's paper was submitted to the National Academy of Sciences, his ideas had made little inroad on actual marriage practices among Deaf people in the United States. Statistics alone could not quell the debate, however, and professionals would continue to use Deaf-Deaf marriages as an argument for limiting Deaf peoples' right to associate with one another.[40]

Deaf people harnessed contemporary standards of love, domesticity, and gender to defend Deaf-Deaf marriages. Deaf people surrounded their marriages with trappings of normalcy, countered calls for marriages influenced by hereditary considerations with a defense of romantic love, and participated in traditional gender norms when presenting the Deaf argument to society at large. The tropes and trappings of romantic love were a crucial tool for presenting Deaf-Deaf marriages as ordinary middle-class marriages. Wedding announcements in Deaf community periodicals in the United States and Great Britain emphasized that Deaf-Deaf marriages were no different from the marriages of hearing people. Unlike occasional sensationalistic pieces in the mainstream press, which stressed the "silence" of Deaf wedding ceremonies and the novelty of people marrying "without uttering a word," wedding announcements in Deaf periodicals emphasized standard features of middle-class weddings, usually gushing about "the fragrance of countless blossoms" at ceremonies and the "handsome and valuable presents" received by the happy couple.[41] These announcements were relentlessly ordinary, differing from announcements to be found in any local periodical only in the passing mention of the presence of an interpreter or the use of sign language. More commonly, bare-bones announcements recited names and dates, named family members and friends in attendance, and wished the couple well in their new life together.

By their nature, wedding announcements are particular to a time and place, of interest to a limited, usually local, set of people. Yet, in the transatlantic Deaf world, a wedding in one locality could find new meaning across the Atlantic. The 1890 wedding of Deaf Americans Annabel Powers and Charles Kearney in Decatur, Illinois, was the subject of a full column article in the British *Deaf and Dumb Times*.[42] The

Kearney and Powers wedding served as an illustration of an elite Deaf-Deaf marriage, Kearney being the head of his own school in Indiana. Also of interest is the presence of Phillip Gillett as the interpreter. Gillett, then superintendent of the Illinois school for Deaf people (at that time the largest school for Deaf people in the world) and a later president of an association Bell founded to promote oral education, was well known and respected in the field of Deaf education.[43] The *Times* article shows him interpreting—and thus validating by his presence—a wedding between two Deaf people. It is impossible to know if Gillett's presence was a factor in the *Times* printing this article; Kearney's prominence and his upcoming trip to Europe could easily be the main reasons. Yet, a point is made by Gillett's presence: support for oralism does not necessarily correspond to support of limits on Deaf people's right to marry whom they chose. This argument was anathema to E. Synes Thompson, chair of the prominent British Ealing Street Training College for Teachers of the Deaf and Diffusion of the Pure Oral System. Some months earlier, Thompson had written to the *British Medical Journal* reaffirming the importance of oral education in preventing intermarriage,

> By [pure oralism] alone can the deaf be taught to communicate with and diffuse themselves among the general population, whilst signing makes them clannish, and intimate association naturally leads to intermarriage.[44]

Thompson's argument reads almost as a synopsis of Bell's *Memoir*, illustrating that both ideas for and against Deaf-Deaf marriage were transferable to debates conducted in other national settings.

Some Deaf people evoked contemporary standards of romantic love in response to exhortations to reject marriage partners for hereditary reasons. According to this argument, "true love. . . based on mutual respect and sympathy" was the basis of Deaf-Deaf marriages. Since Deaf couples enjoyed "genuine sympathy with each other . . . [and] an appreciation of each others desires and feelings," Deaf-Deaf marriages were invariably happier than Deaf-hearing marriages.[45] The ideal of "true love" as a necessary precursor to marital bliss drew upon contemporary notions of romantic love, setting a clear contrast to marriage on hereditary grounds. However, there is an additional element to this argument—the contention that Deaf people felt a "mutual respect and sympathy" for one another because they shared a common experience of being Deaf, an experience Deaf-hearing marriages could not match. Deaf people adopted social conventions of romantic love, adding to them in order to match their unique social circumstances. In doing so, they claimed a privileged status for Deaf-Deaf marriages.

While not all Deaf people necessarily agreed that Deaf-hearing relationships were inferior to Deaf-Deaf ones, the majority held to the position that the choice of a marriage partner was ultimately an individual choice. An 1892 debate at the largely middle-class Chicago Pas-a-Pas Deaf club on the topic of whether women should be allowed to propose marriage featured two teams mixed in gender and marital status. One single female judge and one single female debater participated, as did two married women. Even after the debate concluded, one debater, George Dougherty, "took the stand in favor of women's rights" and declared the right of women to choose their own partner vital to their happiness, as it would allow them to "select the right kind of partner, instead of 'waiting to be asked.'"[46] Dougherty's comments illustrate the primacy of a belief in individual autonomy in domestic affairs assumed by Deaf people in the United States and Britain in this period. As Deaf Briton Francis Maginn asserted at an 1890 London conference of British associations,

> Should the Government dare to interfere with our private domestic life, let us rise to a man and protest—and also let us get up a monster demonstration and march four deep to Trafalgar Square.

Maginn assured his audience members Parliament would, upon seeing such a demonstration, "speedily turn in and erase the words out of the [hypothetical] Bill forbidding the intermarriages."[47] Maginn's comments illustrate an important aspect of the marriage debate: that Deaf men reserved for themselves the right to represent Deaf people on the marriage debate in the general public sphere. As the participants of the Pas-a-Pas club debate show, local clubs and associations afforded Deaf women a space to articulate their thoughts on issues of the day, including marriage. But when it came to public discussion of the marriage issue, it was almost always Deaf men who took the lead.[48] One Deaf American was astonished to greet his old primary school classmate from Connecticut, Matilda Freeman, who was living in Paris and working alongside her husband to organize the 1889 Paris Congress.[49] Likely able to use French Sign Language, American Sign Language, English, and French, Freeman's skills must have proved invaluable in organizing the Congress. Yet her views on the events of the day are not noted in print, nor did there seem to be any Deaf women present at Congress sessions.[50] Freeman's absence illustrates the distinction made between local Deaf events and events which could filter out to the larger public sphere. The Deaf response to the marriage debate may have been the performance of gendered Deaf identities for public consumption—a way of asserting traditional middle-class gender identity in a public debate that attempted to reduce both Deaf men and women to the status of hereditary disorders on the national body.[51] The predominantly Deaf male response to the marriage debate indicates Deaf men knew their citizenship was tied to their masculinity, and if the latter was weakened in public, the former would be also.

While disseminating his message among scientists ensured widespread publicity and scientific weight for his claims, Bell also needed the support of those who worked most closely with Deaf children and Deaf people—educators, ministers, and social workers. Bell had sent advance notice of his 1883 National Academy address to selected educators, and a number were skeptical of his conclusions, publicly stating so after the *Memoir* was published.[52] British educators and missioners were also skeptical of Bell's ideas when they reached Britain, again referring to their own statistics and personal knowledge.[53] Traditionally, fault lines that divided the profession on both sides of the Atlantic tended to be between those emphasizing oral instruction versus those who saw a place for signed languages in Deaf education. The marriage debate did not always follow this division. Most oralists saw the prevention of Deaf-Deaf marriages as an integral part of their overall ideology that Deaf people ought to assimilate into larger society on an individual basis. However, even some educators who supported the use of sign language could agree with this part of the oralist agenda. Educators of sharply different methodological dispositions, including Bell and one of his foremost antagonists, Edward Miner Gallaudet, the president of the National Deaf-Mute College, found common ground in their belief that the good of society necessitated that Deaf people suppress their desire to marry one another.

This tension between individual autonomy and the perceived needs of national societies would erupt on the pages of *Science* in an exchange of letters among the leading figures in American Deaf education in the autumn and winter of 1890 and 1891. Bell started the debate by expressing alarm that, seven years after his *Memoir*, "many of the most prominent teachers of the deaf in America advocate the intermarriage of deaf-mutes."[54] Phillip Gillett, a man who had worked in the field longer than any other American teacher (and the man who presided over the marriage of Kearney and Powers), was singled out as one of these leading educators. Throwing down the gauntlet to Gillett, Bell laid out a case of a young Deaf man from a large multigenerational Deaf family who wished to marry a woman who was congenitally deaf, but had no Deaf family members.

> The teacher of the young lady has been consulted, and she feels her responsibility deeply. Her heart is with the young couple and she desires their happiness, and yet her judgment is opposed to such a union.[55]

Bell stressed that this was not a hypothetical scenario. "The parties are engaged, but the marriage has not yet been consummated," and Gillett's opinion "would have weight with the young couple."[56]

Gillett saw nothing wrong with the marriage "if the parties most interested believe, after reflection, that their own happiness will be promoted thereby." Gillett attacked full out the idea of regulating marriages for any reason, seeing marriage as having "a holier and higher" purpose than "to produce human animals." Gillett's argument paralleled Deaf people's arguments in defense of their right to lead normal lives. The chance of having Deaf children was of little consequence to Gillett, who reasoned that deafness was neither "a crime, or a disgrace, or entailed suffering." Rather it was little more than a "serious inconvenience," comparable to baldness or near-sightedness.[57] This defense of the moral innocence of deafness echoed a statement made by Francis Maginn in London earlier that year: "Why should *one* affliction, and that one innocent in its moral character, be singled out for coercive treatment?"[58] In later letters, Gillett noted the progressive educational institutions of the United States had given Deaf people the chance to "stand upon the same plane as others, and [they] must provide for themselves as others do." Any attempt to restrict their liberties, to treat them "as a special class who are to be looked after by others" would be a disavowal of the goals of independence and self-sufficiency that both Gillett and Deaf people held dear.[59] Gillett's response was a full-throated statement of the importance of individual autonomy and the unassailable sanctity of the matrimonial relationship.

Bell and Gallaudet, the latter declaring deafness was indeed "a grave misfortune," saw the fundamental question in the debate as the effect of Deaf-Deaf marriages on society as a whole. First above all, "the interest of the family must take precedence over the interest of the individual," Gallaudet declared, "for it is the family, and not the individual, that constitutes the basic unit of society."[60] Bell emphasized that while the percentage of Deaf children resulting from Deaf-Deaf marriages was small, the cost to the state of educating these people ran to over $1 million a year.[61] Being a good citizen meant looking out for the public body as a whole beyond individual interests. Gallaudet and other educators exhorted Deaf people to choose celibacy if there was any risk of their passing on their deafness to their offspring. "I would rank high in my esteem a deaf person [with deaf family members] who lived single," Gallaudet wrote in *Science*.[62] More common were recommendations to marry hearing people, usually for the sake of the children. At a national conference in Britain, a Mr. J. Dent, likely a hearing British missioner, reminded the Deaf leaders in his audience, "You must not think only of your own convenience and happiness in this matter, but remember the children."[63] While Bell did not explicitly urge celibacy in his address to the students of the National Deaf-Mute College, he did urge his audience "to remember that children follow marriage, and I am sure that there is no one among the deaf who desires to have his affliction handed down to his children."[64] Children meant the future of the nation, and Deaf people were called upon to renounce individual wants for the national well-being.

While Deaf leaders adamantly opposed legislative restrictions, some publicly conceded that Deaf children were not desirable outcomes of Deaf-Deaf marriages. Deaf American Olof Hanson considered it "self-evident" that marriages which could lead to Deaf children "should be avoided."[65] A summary of the proceedings of the 1890 Conference of Adult Deaf and Dumb Societies in London conceded headmasters of schools should instruct Deaf pupils to "exercise more caution" when choosing marriage partners, "especially where congenitality exists in both the families."[66] Deaf American S.G. Davidson told the delegates to the Paris Congress that Deaf people "should not marry" if they were liable to have Deaf children.[67] What comes through here is the ambiguity of these statements. Individual autonomy in private affairs did not necessarily mean support for Deaf parents having Deaf children, but neither did it mean formal restrictions on choice of marriage partners. By declaring something "self-evident," Hanson was saying that individual Deaf people could be counted on to make the right decision, without others interfering. The use of the word *should* also implies room for maneuver. Deaf men walked a fine line in public, crafting a compromise between larger society's understanding of deafness as an undesirable disability and their own stance that Deaf people ought to be entitled to what Armour called "the liberty of the subject," the same degree of rights and freedoms as that of any other individual.[68] With their rights potentially at stake, Deaf leaders sought a middle ground, refusing to cede individual rights, yet rejecting any attempt to publicly defend the right to have Deaf children.

The danger of children was foremost on the minds of educators, but the Deaf response presented to the public centered on marriages and on hearing children, with little comment on or condemnation of Deaf children.[69] However, when a historian looks at the personal lives of specific individuals, public statements are problematicized, and individuals act with a complexity belied by their public discourse. The case of George Sanders and Lucy Swett is an interesting study of how the Deaf-Deaf marriage debate intimately influenced the lives of two generations of Deaf people and that of Bell himself. Especially in light of the sign language/oral debate of the time, this story reveals the instability of essentialized categories of dominance and resistance.

Ties of finance and friendship connected the Sanders and Bell families. George Sanders's hearing father, Thomas Sanders, a prosperous New England leather merchant, was the first investor in Bell's acoustic experiments, which would eventually lead to the invention of the telephone. The two had met in 1872 when Thomas Sanders asked Bell to teach his five-year-old Deaf son, George. Bell eventually moved to the third floor of Thomas Sanders's mother's home in Salem, Massachusetts, spending the next three years teaching young "Georgie" in the mornings and spending evenings conducting his experiments.[70] Even after the invention of the telephone and his newfound fame, Bell continued to keep close contact with the Sanders family and George in particular. In 1882, fifteen-year-old George Sanders moved to Washington D.C., to continue his studies at the National Deaf-Mute College.[71] Bell reassured Thomas Sanders that his son's contacts would be mostly with "semi-mutes not congenital deaf-mutes" and they would converse in English, not in sign language.[72] Despite Bell's and his parent's intentions, George Sanders's time at the college enabled him to make the acquaintance of Deaf people with whom he would come into contact for the rest of his life.[73]

In his twenties, George Sanders met and fell in love with Lucy Swett, a young Deaf woman from a multigenerational Deaf family influential in the New England Deaf community. Lucy Swett's Deaf great-uncle, Thomas Brown, was the organizer of the first large-scale gathering of American Deaf people in 1850.[74] Her Deaf father, William Swett, was the copublisher of a Boston-based Deaf periodical, the *Deaf-Mute's Friend*, and founder or leader of several Boston-based social, literary, and educational institutions for Deaf people.[75] All of Lucy's aunts and uncles on both sides were Deaf, as was her paternal grandmother, her sister, and a number of cousins.[76] Of Lucy's congenital deafness, there was no doubt.

Nor was there much doubt of George Sanders's love for her. When George's grandmother—she who had hosted Bell in his early inventing days—died, Bell was startled to meet Lucy Swett exiting his old lodgings as he entered to pay his respects. After George had "pleaded with his mother," Lucy was introduced to George's grandmother before she died. George Sanders was unabashed in declaring his interest in Lucy. He visited Lucy at her home soon after the funeral and, with Bell and others watching, "kissed his lady love before them all."[77] Bell had earlier tried to head off their courtship, sending one of his research assistants to investigate the Swett's family deafness and personally passing this information on to George Sanders's mother.[78]

While George Sanders was determined to marry whom he chose, Lucy Swett agonized over the disapproval of Bell and the elder Sanders. "I am keenly alive to my unfortunate position," Swett wrote George's mother of her hereditary deafness, "yet am utterly helpless." Swett's letter to Mrs. Sanders shows the very real passions the marriage debate instilled in one facing not an abstract discourse but a concrete reality.

The letter is a cascade of emotion, alternating between longing, despair, and defiance. "One thing I know—it would break my heart to have a deaf child," Lucy said, defiantly adding, "I should blame very much those persons who agitated the question of hereditary deafness" for this, an allusion to contemporary beliefs that dramatic impressions on a pregnant woman could have a formative impact on fetal development. While acknowledging she understood the family's resistance to their marriage, Swett penned fierce statements of love for her fiancé and described her hopes of one day being in a "dear sweet house of my own."[79]

The courtship of George Sanders and Lucy Swett brought the Deaf-Deaf marriage debate to Bell in a way he had not anticipated. Outside the public sphere, away from debates at congresses and papers presented at scientific societies, Bell was forced to confront fundamental matters of life and love among those he considered closest to himself. Bell felt a special bond with George Sanders. Not only had Bell acted as a teacher and mentor to George for nearly two decades, but he would also later provide George with a job at his new Volta Bureau and, after Thomas Sanders's fortune dribbled away in failed investments, send work and considerable financial support to George Sanders's printing business in Philadelphia.[80] Despite telling his wife "I will not venture to advise him," a close reading of the *Science* debate shows Bell felt very much involved in the couple's decision. For the case Bell presented to Gillett in the pages of *Science*—that of the "young man" with a congenitally deaf family about to marry a young Deaf woman from a hearing family—was the case of George Sanders and Lucy Swett, their genders reversed, but most other details of their lives otherwise the same. The "teacher of the young lady," who felt "her" responsibility deeply and whose heart was with the young couple was none other than Bell himself. [81]

Feeling "desolate and alone" in his hotel after the elder Mrs. Sanders's funeral, Bell wrote a melancholy letter to his wife. Apologizing for his long absences and feeling himself "a dead weight on your young life," Bell reminisced on their early courtship and ruminated on the nature of love. Writing of George and Lucy, Bell noted, "I'm afraid the attachment has become too strong for prudence to have sway. They will surely marry—but what then? Will lovers ever consider the good of those that will come after them?" While Bell admired the young man, noting "he has become a manly fellow and everyone likes him," he felt obligated to admonish Sanders's courtship as improper for hereditary reasons: "George chooses danger to his offspring—for her love." Immediately following this sentence, Bell concludes, "Yet I can understand it too."[82] Proud of the young man George had become, unable casually to dismiss the love of two young people he knew well, Bell acknowledged the inevitability of their union. When the political became personal, Bell acquiesced to a central argument put forth by Deaf people against his ideas in the marriage debate: individual happiness ought to govern over hereditary principles, even when deafness could result.

George and Lucy Sanders would marry and have two children, Dorothy Bell and Margaret. Both were deaf. Lucy Sanders took a pragmatic approach to the lives of her two Deaf children.[83] Sanders spoke orally to her daughters and sent them to public schools with hearing children, as she herself had been educated. She refused, however, to emphasize pure oralism, preferring instead to "[treat] them as I would any hearing child" and let them develop "naturally" in all areas of life, from education to speaking to interacting with other people. Alexander Graham Bell seems likely to have influenced the education of Dorothy Bell and Margaret. The children were put in public schools and their parents also apparently de-emphasized sign language in favor of spoken English.[84] For his part, Bell learned to accept a congenitally deaf intermarriage and its Deaf offspring in a family of close personal acquaintance. After Thomas Sanders's death in 1911, Bell sympathized in a letter to George Sanders: "It must be a great comfort to you to have your wife and children with you now. You are not alone." Expressing the wish to one day see the two girls, Bell asked George to "please give my best love to Lucy and the children and believe me," Bell concluded his letter, "your affectionate friend."[85] It is at the realm of the intimate that complexities emerge belying traditional dichotomies of domination and resistance so prevalent in public discourses.[86]

George and Lucy Sanders were hardly typical of members of the nineteenth-century Deaf community. Most Deaf people were not given private instruction by the era's most noted oral educator, nor were they descended from a prestigious multigenerational Deaf family. Nevertheless, their story illustrates the larger history of the Deaf community in this time and place. Foremost in the Deaf argument was a pragmatic ideology that emphasized Deaf people's right to exist both as autonomous individuals in larger society and within their own communities. While educating their daughters in hearing schools, the Sanders were active in their local Deaf church and renowned for their hospitality to local, national, and international Deaf visitors at their cozy middle-class home in Mt. Airy, Pennsylvania.[87] Dorothy Bell, widowed by World War I, opened a roadside café near Valley Forge to cater to thirsty participants in the modern phenomenon of the recreational automobile trip.[88] After passing the requisite Civil Service examinations, Margaret worked for a time in the Signal Service Corps as a typist.[89] Both continued to participate in the Philadelphia Deaf community while carving their own paths in rapidly modernizing twentieth-century America.

The Deaf argument in the marriage debate was shaped in response to hearing constructions of Deaf people, which were themselves formed by ideas put forth in the public sphere by educators, scientists, and Deaf people. Tracing the ultimate influence of one or another strand of this discourse to specific entities may be an interesting exercise, but what is most compelling is to see how these constructions served as maps—and as blinders—on Deaf and hearing lives. While Deaf people were successful in preventing any formal restrictions being placed on their matrimonial practices, the first decades of the twentieth century would see the rise of a full-blown international eugenics movement, which would sterilize and kill such unwanted burdens to the nation. George and Lucy Sanders married whom they loved regardless of eugenic considerations. Facing a more invasive climate and hoping to avert unwanted attention, twentieth-century Deaf leaders would adopt eugenic considerations and counsel hereditarily Deaf people not to marry one another.[90] While Deaf political discourses could hold sway for a period, the weight of larger discourses would eventually serve to prearrange the boundaries of choices and actions of individual Deaf people. Deaf people retained their right to choose their marriage partners based on true love and sympathy, but the anti-marriage discourse was likely successful in instilling genetic considerations as criteria of love among some members of the Deaf community.

More than a story of a single minority, the Deaf-Deaf marriage debate sheds light on early discursive skirmishes embarked upon by the emerging transnational science of eugenics. While Deaf people would successfully maintain their traditional marriage rights, the promotion of the needs of society over that of the individual, justified by seemingly objective scientific standards, would prove to be successful when applied to other stigmatized groups in the twentieth century. The Deaf-Deaf marriage debate also reveals the strengths and limits of transnational minority discourses. Existing throughout the world without regard to national boundaries, Deaf people utilized similarities in their experiences to form potent counterarguments to a transnational discourse. Politically scattered into small communities and facing intense pressures from their surrounding national societies, Deaf people could not always sustain their arguments over generations against an aggregation of discourses and ideas arrayed against them. Facing new ideologies, Deaf people devised new counterarguments, consistently interjecting themselves in discourses denigrating their claims to full citizenship. Deaf people did not stay in the latter half of a binary relationship of domination and resistance; they negotiated multiple and changing relations to power at different times and places. As the Sanders family entered the twentieth century, the next generation of Deaf people would continue to present themselves as healthy, contributing members of their nation-states.

NOTES

1. Italics in original. Alexander Graham Bell, *Memoir Upon the Formation of a Deaf Variety of the Human Race* (Washington, D.C.: Volta Bureau, 1883), 4.

2. "consumed . . ." in "The November Meeting of the National Academy of Sciences," *Science* (23 November 1883): 1. "in the formation . . ." Bell, *Memoir*, 4.

3. Standard practice in the field of Deaf Studies distinguishes between "Deaf" people, who are members of a sign language-using community, and "deaf" people, who cannot hear but do not use their national signed language or identify with a community of Deaf people. Earlier practice among professional historians writing histories of Deaf people has been to refer to all people who do not hear as "deaf," leaving it up to the reader to determine cultural boundaries. This seems to privilege a noncultural state of being when it can be argued the opposite is true for works that examine the signing Deaf community. The Deaf people studied in this essay are, for the most part, advocates

for their national signed languages and Deaf communities. This essay will refer to them as Deaf, not deaf.

4. David Buxton, *On the Marriage and Intermarriage of the Deaf and Dumb* (Liverpool: W. Pearnall and Co., 1857), 16.

5. Discussion of the causes of "deaf-mutism" had appeared in the very first issue of the widely circulated journal of Deaf education, the *American Annals of the Deaf and Dumb*, in 1848. William Wolcott Turner, "Causes of Deafness," *American Annals of the Deaf and Dumb* 1 (1848): 25–32. Bell attributed the theory of a "deaf-mute race" to Turner, somewhat disingenuously claiming he was merely passing on the latter's ideas with his own statistical research. This ignores the fact that Turner was speaking to a narrow audience of educators while Bell ensured his ideas received maximum exposure among scientists and the general public. Alexander Graham Bell, "Professor A. Graham Bell's Studies of the Deaf," *Science* 16:396 (5 September 1890): 135–36.

6. "Proceedings on the Section of Anthropology," *Science* 4:87 (3 October 1884): 343–46; "The National Academy of Sciences," *Scientific American* (21 November 1885): 320; "National Academy of Sciences—Washington Meeting." *Scientific American* (22 May 1886): 328.

7. The phrase "common strategy of response" is adopted from Ann Laura Stoler's call for scholars to look for "common strategies of rule" when undertaking comparative projects. A central element of Stoler's comparative analysis is a suggestion that scholars ought to see if similar patterns of power emerge in different local contexts. "Tense and Tender Ties: The Politics of Comparison in North American History and (Post) Colonial Studies," *Journal of American History* 88:3 (December 2001): 847.

8. Edward Allen Fay, *Marriages of the Deaf in America* (Washington, D.C.: Volta Bureau, 1893), 18–19.

9. Historian Nancy F. Cott observes that late-nineteenth-century American public policy promoted the role of men as husbands and heads of households, a role rooted in the "presumed conjunction of marriage, property-owning, household headship, and male citizenship." Nancy F. Cott, *Public Vows: A History of Marriage and the Nation* (Cambridge, Mass.: Harvard University Press, 1992), 122. Leonore Davidoff and her co-authors sketch a similar position for men in British society, noting "personal authority within the family can be linked to much broader relationships of power" dealing with individual identity and autonomy. Leonore Davidoff, Megan Doolittle, Janet Fink, and Katherine Holden, *The Family Story: Blood, Contract, and Intimacy, 1830–1860.* (New York: Longman, 1999), 137.

10. R. A. Soloway, Demography and Degeneration, (Chapel Hill: University of North Carolina Press), 21.

11. For a comparative look at British and American eugenics, see Marouf Arif Hasian, Jr., The Rhetoric of Eugenics and Daniel Kevles, In the Name of Eugenics: Genetics and the Uses of Human Heredity New York: Knopf, 1985. For international contact among scientists, see Martin S. Pernick, The Black Stork: Eugenics and the Birth of "Defective" Babies in American Medicine and Motion Pictures, (New York: Oxford University Press, 1996), 22-23; Randall Hansen and Desmond King, "Eugenic Ideas, Political Interests, and Policy Variance: Immigration and Sterilization Policy in Britain and the U.S.," World Politics 53 (January 2001) 240.

12. The chairmanship of the committee was then offered to David Starr Jordan. Robert V. Bruce, Bell: Alexander Graham Bell and the Conquest of Solitude (Boston: Little, Brown, and Company, 1973), 417-418.

13. Donald K. Pickens states that the nineteenth century ended in a superseding of traditional liberal beliefs in human rationality with a politics of the collective based on genetic considerations. Genetically inferior individuals were essentialized into potential problems for the national body. Donald K. Pickens, *Eugenics and the Progressives* (Nashville, Tenn.: Vanderbilt University Press, 1968), 45.

14. Hawkins, *Social Darwinism in European and American Thought, 1860–1945.* (New York: Cambridge University Press, 1997), 219–22.

15. Pickens, *Eugenics and the Progressives*, 44–45; Martin Pernick, "Public Health Then and Now," *American Journal of Public Health* 87, no. 11 (November 1997): 1769. For an overview of British thought at the turn of the century, see Hawkins, *Social Darwinism*, 222–31 and Soloway, *Demography*, 2–17. Pickens offers a good summary of American hereditarianism thought in the nineteenth century. Pickens, *Eugenics and the Progressives*, 37–46.

16. For the impact of hereditarian and eugenic thought on women's movements in the United States and Britain at the turn of the century, see Linda Gordon, *Woman's Body, Woman's Right: A Social History of Birth Control in America* (New York: Grossman Publishers, 1976), 116–58 and Richard Allen Soloway, *Birth Control and the Population Question, 1877–1930* (Chapel Hill: University of North Carolina Press, 1982), 133–55. For a comparative look, see Marouf Arif Hasian, Jr., *The Rhetoric of Eugenics in Anglo-American Thought*, 72–88.

17. These laws typically restricted the marriages of those deemed a threat to the national body politic, usually of the feebleminded, but also "the insane,

syphilitic, alcoholic, epileptic and certain types of criminals." Connecticut was the first, followed by Kansas (1903); New Jersey, Ohio, Michigan, and Indiana (1905); Indiana (1907); and California. Pickens, *Eugenics and the Progressives*, 88. By 1915, thirteen states had sterilization laws as well. Pickens, *Eugenics and the Progressives*, 90.

18. Bell, *Memoir*, 45.

19. Ibid., 46.

20. By placing Deaf children in public schools, Bell aimed to eliminate the acculturation into the Deaf community that went on at Deaf schools. This would presumably foster their assimilation into their national majority culture. "The November Meeting of the National Academy of Sciences," *Science* 23 (November 1883): 1; Bell, *Memoir*, 47.

21. The *Oxford English Dictionary* offers three definitions of "intermarry" or "intermarriage." The first and most generic is a hyphenated form of the word (inter-marriage), which simply means the marriage of two people. This was superseded by the more general term "marriage;" the "inter" generally being used only in legal documents after 1800. A second nineteenth-century use of this word meant "marriage between persons (or interbreeding between animals) nearly related; consanguineous marriage or breeding," [intermarriage] or, more simply, "to marry with each other" [intermarry]. This term emerged in 1843 and 1855 to refer to marriages within a small community, such as a village, to the extent that nearly everyone was related to one another. The word was first used to refer to the interbreeding of animals in 1882. A subset of the first definition is "the marriage of persons of different families, castes, tribes, nations, or societies, as establishing a connexion between such families, etc." This definition was used most often with reference to royal-commoner marriages in the early modern period. Over time, and especially from 1880 to 1900, "intermarriage" gained its colonial connotations. Definitions and dates of usage presented below are all present in the 1989 second edition of the *Oxford English Dictionary*, online at www.oed.com.

22. For more on these communities, see Harlan Lane, Richard Pillard, and Mary French, "Origins of the American Deaf World: Assimilating and Differentiating Societies and Their Relation to Genetic Patterning," *Sign Language Studies* 1, no. 1 (Fall 2000): 17–44.

23. Alexander Graham Bell, *Facts and Opinions Related to the deaf from America* (London: Spottiswoode and Co., 1888); W. G. Jenkins, "The Scientific Testimony of 'Facts and Opinions,'" *Science* 16, no. 395 (August 15, 1889): 85–86.

24. Susan Burch, *Signs of Resistance: American Deaf Cultural History, 1900–1942* (New York: New York University Press, 2002), 140, 205; John Vickrey Van Cleve and Barry A. Crouch, *A Place of Their Own: Creating the Deaf Community in America* (Washington, D.C.: Gallaudet University Press, 1989), 148–49; Lane, *When the Mind Hears*, 358–61.

25. "Developing a Deaf Race- The New York Tribune," *Current Literature* (April 1889): 292.

26. "Reminisces of the Deaf and Dumb Congress in Paris," *Deaf and Dumb Times* (November 1890), 83.

27. Other scholars have noted that the similarity of ideas in different national contexts can indicate the influence of preexisting ideas from extranational sources. For other studies of this form of transnational contact, see S. Ilan Troen, "Frontier Myths and Their Application in America and Israel: A Transnational Perspective," *Journal of American History* 86, no. 3 (December 1999): 1209–230; Rob Kroes, "America and the European Sense of History," *Journal of American History* 86, no. 3 (December 1999): 1135–155.

28. "Reminisces of the Deaf and Dumb Congress in Paris," *Deaf and Dumb Times* (November 1890): 83.

29. This is adapted from the term "tactical mobility of concepts" from Stoler, "Tense and Tender Ties," 837.

30. In fact, Clerc apparently faced some resistance from Thomas Gallaudet to the idea of his marrying a Deaf woman (as related in Lane, *When the Mind Hears*, 263). Lane bases his account on "Retirement of Mr. Clerc," *American Annals of the Deaf* 10 (1858): 181–83.

31. Van Cleve and Crouch also note challenges to Bell's statistics and the happiness argument, locating them as originating from educators of Deaf people. Where these arguments originated is less important to this essay than the fact that both Deaf people and like-minded educators used them consistently throughout the debate. Van Cleve and Crouch, *A Place of Their Own*, 149.

32. "National Conference," *Deaf and Dumb Times*, 109.

33. Ibid., 111.

34. "Most convincing . . ." Ibid., p. 111. In 1890, the superintendent of the Colorado School for the Deaf and Blind reported none of their congenitally Deaf "pupils had Deaf parents. *Report of the Board of Trustees of the Institution for the Education of the Deaf and the Blind* (Denver: Collier and Cleaveland, 1890). For an account of other statistics collected by educators, see Van Cleve and Crouch, *A Place of their Own*, 149.

35. Edward Allen Fay, "Marriage Records," *Deaf-Mutes Journal* 18, no. 45 (November 7, 1889): 4.

36. Fay received back 4,471 surveys in all. Edward Allen Fay, "An Inquiry Concerning the Results of Marriages of the Deaf in America," *American Annals of the Deaf* 41 (1896): 79.

37. Fay, *Marriages of the Deaf in America* (Washington, D.C.: Volta Bureau, 1893), 11.

38. Ibid., 18–19.

39. According to Fay's survey, 72.5 percent of Deaf people married other Deaf people. Fay, *Marriages*, 124.

40. Unlike Bell, some American oralists would later openly call for government measures to prohibit marriages of people "liable to transmit defects to their offspring." Mary S. Garrett, "The State of the Case," *National Educational Association: Journal of Proceedings and Addresses of the Thirty-Ninth Annual Meeting* (Washington, D.C., 1900): 663, as quoted in Douglas Baynton, *Forbidden Signs: American Culture and the Campaign Against Sign Language* (Chicago: University of Chicago Press, 1996), 30.

41. An example of sensationalistic accounts in the mainstream press can be found in an 1888 Chicago Tribune article depicting a Deaf-Deaf wedding where "nobody spoke a word. All were deaf and dumb." The article goes on to detail the "fluttering hands" and "silent gossip" of the guests, ending with the tears of an "old colored attendant, who was crying at the door. 'It's the prettiest thing I ever saw,' said he." Anonymous, "Wooed and Wed by Signs-Chicago Tribune" *Current Literature* December 1888, 500. See also "A Remarkable Wedding," *St. Louis Globe-Democrat*, reprinted in the Deaf-Mute's Journal (November 18, 1880): 2. "Blossoms" and "presents" can be found in "Foreign News: America," *Deaf and Dumb Times* 2, no. 4 (September 1890): 53–54.

42. "Foreign News: America," *Deaf and Dumb Times* 2, no. 4 (September 1890): 53–54.

43. An 1894 article in the British *Quarterly Review of Deaf-Mute Education*, which reviewed Gillett's career, shows he was well known in Great Britain as someone who saw benefit in the oral method. Gillett at the time advocated the "Eclectic System," a system somewhere between the pure oral and combined systems. 'J.F.' "Party Spirit v. Philanthropy," *Quarterly Review of Deaf-Mute Education* 3 (October 1894): 356–59.

44. E. Symes Thompson, "The Deaf and Dumb," *British Medical Journal* (26 October 1889): 950. For its part, the *British Medical Journal* noted, "The

feeling of the medical profession has been again and again strongly expressed as to the superiority of speech over signs." "The Deaf and Dumb," *British Medical Journal* (28 September 1889): 727.

45. "True love," in an address given by Deaf American Thomas Davidson "Reminisces of the Deaf and Dumb Congress in Paris," *Deaf and Dumb Times* (November 1890): 82; "genuine sympathy," in comments by Agnew, a British Deaf man. "National Conference of Adult Deaf and Dumb Missions and Associations," *Deaf and Dumb Times* (March 1890): 111. For an American publication stating that Deaf-Deaf marriages were happier than Deaf-hearing marriages, see Hiram Phelps Arms, *The Intermarriage of the Deaf: Its Mental, Moral, and Social Tendencies* (Philadelphia: Burk and McFetridge, 1887). Phelps makes this point repeatedly throughout the text.

46. Implicit in this is an understanding of the autonomy of Deaf women, which, while not unprecedented by the standard of the day, was certainly progressive. "World's Fair City," *Deaf-Mutes Journal* (21 April 1892): 3.

47. "National Conference," *Deaf and Dumb Times*, (March 1890): 111.

48. The extent of Deaf women's participation in Deaf community debates likely varied according to local and national contexts. This dual public spheres strategy was not unique to Deaf people. Elsa Barkley Brown shows Reconstruction-era African American women in Richmond, Virginia, were active participants in an "internal" African American public sphere, which Brown sets in contrast to what she calls the "external" public sphere, that of white American society. In the external arena, black women were disenfranchised, as were white women. Likewise, Deaf women may have participated in the Deaf public sphere, but kept their presence invisible when it came time to present the opinions of the community to outsiders. Elsa Barkley Brown, "To Catch the Vision of Freedom," *African American Women and the Vote, 1937–1965* (Amherst: University of Massachusetts Press, 1997).

49. Amos Draper, "Notes on the Meeting of the Deaf in Paris," *American Annals of the Deaf* 35 (1890): 33.

50. Draper called the absence of women at the Paris Congress one of its "marked and not at all agreeable features." Amos Draper, "Report to the President," *Annual Report of the Columbia Institution for Deaf Mutes, 1889* (Washington, D.C., 1889), 34. A British report on the Congress, however, indicates "several foreign ladies" in the audience on the second day of Congress proceedings. "Reminiscences of the Paris International Deaf and Dumb Congress," *Deaf and Dumb Times* (October 1890): 69.

51. Joan Scott notes politically charged events provoke "the suppression of ambiguities and opposite elements [of gender identity] in order to assure . . . coherence and common understanding." Scott, "Gender: A Useful Category of Historical Analysis," *American Historical Review* 91, no. 5 (December 1986): 1073.

52. "Advance notice," letter from Alexander Graham Bell to Edward M. Gallaudet, 11 October, 1880 [sic, letter is dated 1883] (Series: General Correspondence, Folder: Edward M. Gallaudet, 1877–1902. undated), the Alexander Graham Bell Family Papers at the Library of Congress, 1862–1939, accessed via the American Memory Project, http://memory.loc.gov/ammem/bell-html/bellhome.html [hereafter, Bell Papers]; letter from Alexander Graham Bell to Sarah Fuller, 13 October, 1883. (Series: Subject File, Folder: The Deaf, Fuller, Sarah, General Correspondence, 1883–1890), Bell Papers. Not only did heads of schools for Deaf people question Bell's statistics, but some even agreed that Deaf-Deaf marriages were happier than Deaf-hearing marriages. Bell, *Facts and Opinions Related to the deaf from America* (London: Spottiswoode and Co., 1888); Van Cleve and Crouch, *A Place of Their Own*, 149.

53. "Correspondence," *Deaf and Dumb Times* 3, no. 1 (August 1889): 27; "National Conference of Adult Deaf and Dumb Missions and Associations," *Deaf and Dumb Times* (March 1890): 109–12.

54. Bell, "Professor A. Graham Bell's Studies of the Deaf," *Science* 16, no. 396 (5 September 1890): 136.

55. Ibid.

56. Ibid.

57. Phillip Gillett. "Deaf-Mutes," *Science* 16, no. 404 (31 October 1890): 248–49.

58. "National Conference," *Deaf and Dumb Times*, 111.

59. Phillip Gillett, "The Intermarriage of the Deaf, and Their Education," *Science* 16, no. 412 (26 December 1890): 353–57.

60. Edward Miner Gallaudet. "The Intermarriage of the Deaf, and Their Education," *Science* 16, no. 408 (28 November 1890): 295–99.

61. Alexander Graham Bell, "Deaf-Mutes," *Science* 16, no. 412 (26 December 1890): 358–59.

62. Gallaudet, "The Intermarriage of the Deaf, and Their Education," 296.

63. "National Conference," *Deaf and Dumb Times*, 110.

64. Alexander Graham Bell, *Marriage: An Address to the Deaf* (Washington, D.C.: Volta Bureau, 1891), 4.

65. Olof Hanson, "The Tendency Among the Deaf to Exclusive Association With One Another," *American Annals of the Deaf* 33 (1888): 30.

66. "Our Notes," *Deaf and Dumb Times* (February 1890): 82.

67. "Reminisces of the Deaf and Dumb Congress in Paris," *Deaf and Dumb Times* (November, 1890): 82.

68. "Reminisces of the Deaf and Dumb Congress in Paris," *Deaf and Dumb Times* (November 1890): 83.

69. The only print mention I have seen of Deaf children by a Deaf author is in Hiriam Phelps Arms, *The Intermarriage of the Deaf*, 34–35. Arms writes, "As to the offspring being deaf and mute like their parents, 'what of that?'" saying Deaf people had the ability to accomplish as much as hearing people if they so desired. More common is a stress on hearing children of Deaf parents. Historian Susan Burch points out a series of articles from the *Silent Worker* from between 1917 and 1929, which emphasized the accomplishments of hearing children of Deaf parents. Burch, *Signs of Resistance*, 145.

70. The chairmanship of the committee was then offered to David Starr Jordan. Robert V. Bruce, *Bell: Alexander Graham Bell and the Conquest of Solitude* (Boston: Little, Brown, 1973), 417–18.

71. Bruce, *Bell*, 399. For Bell's account of his instruction of George Sanders in the 1870s, see Bell's *Upon a Method of Teaching Language to a Congenitally Deaf Child* (Washington, D.C.: Gibson Brothers, Printers, 1891).

72. Alexander Graham Bell to Thomas Sanders, 4 January 1883 (Series: Subject File, Folder: The Deaf, Sanders, George, 1882–1913, undated), Bell Papers.

73. Florence E. Long, "Stray Straws," *Silent Worker* 32, no. 4 (January 1920): 95.

74. Lane, Pillard, and French, "Origins of the American Deaf World," 30–31.

75. Ibid., 35.

76. Ibid., 20.

77. Alexander Graham Bell to Mabel Hubbard Bell, 5 May 1890, (Series: Family Papers, Folder: Mabel Hubbard Bell, Family Correspondence, Alexander Graham Bell, 1890), Bell Papers.

78. Alexander Graham Bell to Mabel Hubbard Bell, 14 October 1889, (Series: Family Papers, Folder: Mabel Hubbard Bell, Family Correspondence, Alexander Graham Bell, May 1889). Bell Papers.

79. Letter from Lucy M. Sanders to Mrs. Sanders, undated. (Series: Subject File, Folder: The Deaf, Sanders, George, 1882–1913, undated), Bell Papers.

80. Bruce, *Bell*, 399–400. Bruce suggests Bell saw George Sanders as a surrogate son, having none of his own. George Sanders worked at the Volta Bureau when it first started up at a salary of $50 a month. His primary occupation at one time was to record marriage announcements from the *Deaf-Mutes Journal*. Letter from John Hitz to Alexander Graham Bell, 17 January 1893. (Series: Subject File, Folder: The Deaf, Sanders, George, 1882–1913, undated), Bell Papers. At one point, Bell's wife complained, "George Sanders costs you $2000.00 a year, and what do you really get from him?" Mabel Hubbard Bell to Alexander Graham Bell, 13 May 1898. (Series: Family Papers, Folder: Mabel Hubbard Bell, Family Correspondence, Alexander Graham Bell, May 1898) Bell Papers.

81. The family history of the "young man" in Bell's letter matches Lucy Swett's in nearly every detail. "A young man (not a deaf-mute) became deaf in childhood while attending public school. He has one brother who is a deaf-mute . . . [his] father was born deaf in one ear and lost the hearing of the other . . . [the father also] had a congenitally deaf brother who married a congenital deaf-mute and had four children (three of them congenital deaf-mutes). The mother of the young man was a congenital deaf-mute, and she also had a brother born deaf. The paternal grandmother . . . was a congenital deaf-mute, and she had a brother who was born deaf. This brother married a congenital deaf-mute, and had one son born deaf. The great-grandfather of this young man . . . was a congenital deaf-mute. Thus deafness has come down to this young man through four successive generations, and he now wants to marry a congenital deaf-mute. The young lady has seven hearing brothers and sisters, and there was no deafness in her ancestry, but she herself is believed by her family to have been born deaf." Bell, "Professor A. Graham Bell's Studies," 136.

82. Bell to Mabel Bell, 5 May 1890. Bell Papers. Also quoted in Bruce, *Bell*, 399.

83. While admitting to "profound shock" upon the discovery of her daughter's deafness, she somewhat incongruously ascribed this to being "so accustomed . . . to the fact that most children of deaf parents are possessed of the sense of hearing." Lucy Sanders, "How My Children Were Educated," *Silent Worker* 23, no. 10 (July 1911): 183.

84. Lucy Sanders claimed, "it was not until the girls were five and seven years old respectively that they learned not *all* people hear and speak, and

they learned the manual alphabet and used signs in a fair way." Sanders, "How My Children Were Educated," 184. This is somewhat surprising, since they had at least one Deaf cousin and were active in their local Deaf church. Charles Dantzer, "Pennsylvania," *Silent Worker* 21, no. 6 (March 1909): 102. Their Deaf cousin was Helena L. Bowden, the daughter of Lucy Sanders's Deaf sister, Persis Bowden. James Reider, "Philadelphia," *Silent Worker* 28, no. 10 (July 1916): 186–87.

85. Alexander Graham Bell to George T. Sanders, 14 August 1911. (Series: Subject File, Folder: The Deaf, Sanders, George, 1882–1913, undated.) Bell Papers.

86. Taking her cue from Foucault, Stoler contends that it is at the level of intimacy and personal relationships that the workings of dominant relations of power are most apparent. See Stoler, "Tense and Tender Ties" for an elaboration of this theme as applied to comparative colonial studies.

87. Henri Gaillard, *Gaillard in Deaf America: A Portrait of the Deaf Community, 1917,* ed. Robert H. Buchanan, (Washington, D.C.: Gallaudet University Press, 2002), 134; Dantzer, "Pennsylvania," 102; Long, "Stray Straws," 95.

88. E. Florence Long, "Laurel Cabin Tea Room," *Silent Worker*, 33, no. 5 (February 1921): 173.

89. Arsene Dozois, "Speaking of a National Magazine," *Silent Worker* 31, no. 5 (February 1919): 77.

90. Burch notes the rhetorical stance taken by late-deafened American Deaf leaders against marriages of congenital Deaf people in the early-twentieth century, saying it had little effect on actual marriage practices. Burch, *Signs of Resistance*, 142–44.

DEAFNESS AND EUGENICS IN THE NAZI ERA

John S. Schuchman

Adolph Hitler became the Chancellor of Germany in January 1933. A short time earlier, the German deaf community, represented by the Reich Union of the Deaf of Germany (REGEDE), produced an educational film *Verkannte Menschen* (Misjudged People).[1] A brief examination of this 1932 motion picture reveals that the purpose of the silent, subtitled, thirty-minute film was to demonstrate to German audiences that deaf people were good Germans—happy, healthy, and capable of employment. To paraphrase the often-cited statement attributed to Gallaudet University President I. King Jordan, the film attempted

---

In the summer of 1998, my colleague, professor Donna Ryan, and I co-chaired the conference "Deaf People in Hitler's Europe, 1933–1945," sponsored by Gallaudet University and the United States Holocaust Memorial Museum. These remarks represent a synthesis of some of the work presented at that conference and recently published by Gallaudet University Press as an anthology, *Deaf People in Hitler's Europe* (Washington, D.C.: Gallaudet University Press, 2002).

to illustrate "that deaf people can do anything except hear." However, as a part of Hitler's *Gleichschaltung* program throughout 1933 and 1934 to coordinate all German institutions with Nazi philosophy, *Verkannte Menschen* was banned.

An examination of the last few minutes of the film reflect the film-maker's effort to demonstrate that deaf people were literate, productive, healthy, and civic-minded Aryans. The film's end also featured the statistic that "90 percent of deaf parents produced children who could hear." This picture of deaf people as healthy Aryans stood in contrast to discussions in the German public after World War I about the need for sterilizing disabled Germans.[2] It is ironic that the deaf scriptwriter for the film, Wilhelm Ballier, was himself active with the deaf Nazis. German Deaf community historian Jochen Muhs has wondered why it was that men who were late-deafened and who spoke well often supported the actions of the Nazi government.[3] In fairness to these men, however, discussions of eugenics always took place within the context of heredity. Even in the United States, the National Association of the Deaf cautioned "congenitally deaf" couples from marriage.[4] The situation in Germany, though, was different. Although the Nazis used the rhetoric of heredity when referring to disabled people, it is clear that, in practice, they viewed all deaf persons as inferior and that "a deaf variety" (to borrow a term from Alexander Graham Bell) had no place in the Nazi worldview of the Aryan race.

Still, it was eugenicists who provided the Nazis with the academic and scientific justification for the party's policy on race. In 1931, for example, University of Munich Eugenics Professor Fritz Lenz asserted that "Hitler is the first politician with truly wide influence who has recognized that the central mission of all politics is race hygiene and who will actively support this mission."[5] Several Holocaust scholars, such as Henry Friedlander (*The Origins of Nazi Genocide: From Euthanasia to the Final Solution* [1995]) and Robert Proctor (*Racial Hygiene: Medicine under the Nazis* [1988]), have described the symbiotic relationship between the eugenicists and the Nazis.[6] The former provided the ideological justification for Nazi political actions, and once in power, the Nazi government provided the scientists with opportunities to implement their philosophy and practices of racial hygiene.

Although not all German eugenicists supported anti-Semitism or "negative" eugenics and sterilization, the advocates who promoted the superiority of Nordic or Germanic people and the exclusion of alien races became dominant under the Nazi regime. The so-called Nordic view of eugenics became racial hygiene. Nazis quickly moved to cleanse Aryan blood of alien and "unfit" Germans. The largest numbers of aliens included Jews and Gypsies. The unfit included physically and mentally disabled Germans.

Hitler's cabinet issued the Law for the Prevention of Offspring with Hereditary Diseases in the summer of 1933, effective January 1, 1934, parallel to the legislation designed to limit or exclude Jews and Gypsies from public life. The new Nazi law, based on an earlier Prussian legislative proposal, allowed for the compulsory sterilization of physically and mentally disabled Germans. Within two years, the Nazis developed a series of laws designed to prevent disabled persons from procreation. These laws focused on sterilization, abortion, and marriage. Subsequent legislation permitted abortions where pregnancy occurred prior to sterilization, even though abortion was generally illegal in Germany. In the fall of 1935, the government enacted the Law for the Protection of the Hereditary Health of the German Nation. Designed to prevent marriages of persons with hereditary diseases, the law required couples to obtain from a public health office a certificate that they were free of hereditary disease.

Implementation of these laws did not require the assistance of the Sturmabteilung (SA), Brown-shirted fascist thugs, or storm troopers.[7] Rather, professionals from such disciplines as medicine, law, education, and social work participated in this physical attack on disabled German citizens, suggesting a widespread acceptance of eugenic beliefs. In *Crying Hands: Eugenics and Deaf People in Nazi Germany*, the late Horst Biesold points out that courses in eugenics, heredity, and race hygiene represented an integral part of the training of educators of deaf children in Germany.[8] When Biesold surveyed approximately 1,200 sterilization survivors in the early 1980s, he found that German teachers and administrators at schools for deaf children actively participated in the identification of deaf children and reported them to the health authorities responsible for implementation of the law. By war's end, a total of between 300,000 and 400,000 disabled persons had been sterilized, according to the best estimates.[9]

The vast majority of sterilizations occurred with persons with a variety of mental disabilities, but Biesold's analysis of the experience of

deaf people suggests that between 16,000 and 17,000 suffered sterilizations.[10] Although deaf people represented only 4 percent of the total
number of sterilized disabled persons, the number reflects a devastating impact on the deaf population of Germany. It has been estimated
that there were 40,000 deaf persons in Germany in 1932. Sixteen thousand sterilizations represent 40 percent of this total. It is ironic that the
number of sterilizations exceeds leading German eugenicists' estimates
of the genetically deaf population. Eugen Fischer, director of the Kaiser
Wilhelm Institute for Anthropology in Berlin, estimated that there
were between 10,000 and 13,000 hereditarily deaf Germans in 1933.[11]
Sixteen thousand sterilizations therefore suggests that the definition of
"hereditarily deafened" was imprecise and that the Health Courts, established to oversee appeals of sterilization orders, operated with a very
broad definition.

The reality is that eugenics science provided a rationale and cover
to individuals who had a pejorative view of disability and deaf people.
The Nazi government produced several propaganda films designed to
win public support of these dangerous views. Holocaust historian
Michael Burleigh has described these films as "intended to criminalize,
degrade, and dehumanize the mentally and physically handicapped so
as to justify compulsorily sterilizing them." In these films, deafness was
linked to "idiocy and feeblemindedness." The term *deaf* never appeared
singly; it always appeared as *deaf and dumb*.[12]

In the Nazi worldview, sterilization was not enough. After World
War I, German authors Alfred Hoche and Karl Binding, in *Authorization
for the Destruction of Life Unworthy of Life* (1920) argued that economic
savings justified the killing of "useless lives." Due to planned programs
of rationing food and medical supplies during World War I, some disabled asylum patients were allowed to die from starvation. Hitler adopted
the "useless lives" view of Hoche and Binding and approved the use of
"euthanasia" against persons with disabilities. Shortly before the outbreak
of World War II in the fall of 1939, the Nazi government organized a
secret program, known as "Operation T4," designed to kill disabled
persons. Unlike the public rhetoric of "useless lives," which usually
meant people labeled as "idiots" and "congenitally crippled," the targets
of the Operation T4 were the weakest disabled citizens—infants, elderly persons in nursing homes, and asylum patients. Both Friedlander
in his "Holocaust Studies and the Deaf Community," published in

Deaf People in Hitler's Europe, and Biesold in *Crying Hands* provide evidence that deaf persons died in the T4 program.

The deaf community had no place in Germany under the Nazis. Medical doctors, lawyers, and educators of deaf children assisted in subjecting deaf Germans to sterilization, abortion, marriage restrictions, and sometimes murder. The German government provided limited assistance to deaf people only when they could contribute to the Nazi claims of racial superiority. The best examples of this occurred prior to the war when the Nazi government sent athletic teams to participate at the world games for the deaf in London (1935) and Stockholm (1939).[13] Education for deaf children was another matter.

In the February 1934 issue of *Blätter fur Taubstummenildung* (Journal of the Education of the Deaf) Kurt Lietz argued in "The Place of the School for the Deaf in the New Reich" that education of deaf children should be comprised of teaching communication techniques and vocational skills. Speech and speechreading instruction should be made available to the limited number of students who could produce intelligible speech. Less-gifted students would receive language instruction through writing, and a large amount of time would be devoted to manual training, with less need for trained deaf educators.

> Such a reorganization would enable schools to get along with considerably reduced budgets, and when the numbers of pupils gradually diminishes, the buildings will be available for other purposes. Even under these circumstances, the cost of the education of the deaf will still remain comparatively high, but it will be more justifiable.
>
> The school for the deaf, therefore, in the New Germany will occupy an entirely different place. The steps to be taken may seem extraordinarily harsh, but they are biologically necessary. The whole revolution draws its strength from the biological foundations of our nation.[14]

Lietz pointed out in the same article that the biological goals of the German nation superceded the societal need to educate deaf children. A German's obligation to the state was to "bear arms" or "bear children." Making no distinction between a hereditarily deaf person or an otherwise normal deaf individual, he concluded that a deaf person could never be a German citizen, merely a German subject.

One may concede that the German eugenicists could not anticipate the abuse that occurred under the Nazi regime, but it is clear that their rhetoric and practices resulted in a milieu of disrespect for disabled people in general, including the deaf population. Without respect for a deaf citizen's human right to pursue a productive and happy life as he or she chooses, discrimination or worse can flourish. The German example in 1933–1945 represents the low point of deaf history, and we need to remind ourselves of it in our daily interactions with deaf individuals and the deaf community.

NOTES

1. For an extensive analysis, see John S. Schuchman, "Misjudged People: The German Deaf Community in 1932," in *Deaf People in Hitler's Europe,* ed. Donna Ryan and John S. Schuchman (Washington, D.C.: Gallaudet University Press, 2002), 98–113.

2. See Jochen Muhs, "Deaf People as Eyewitnesses of National Socialism," in Ryan and Schuchman, eds., *Deaf People,* 81.

3. Ibid., 93–94.

4. National Association of the Deaf, *Proceedings of Thirteenth Convention* (Detroit, Mich.: NAD, 1920).

5. Quoted by Henry Friedlander, "Introduction," in Horst Biesold, *Crying Hands: Eugenics and Deaf People in Nazi Germany* (Washington, D.C.: Gallaudet University Press, 1999), 4.

6. Henry Friedlander, *The Origins of Nazi Genocide: From Euthanasia to the Final Solution* (Chapel Hill: University of North Carolina Press, 1995); Robert Proctor, *Racial Hygiene, Medicine Under the Nazis* (Cambridge, Mass.: Harvard University Press, 1988).

7. However, sterilization was compulsory. If an individual resisted, the local police could and did transport the patient to the hospital for sterilization.

8. Biesold, *Crying Hands.*

9. There are various estimates, but this is the one used by the United States Holocaust Memorial Museum. United States Holocaust Memorial Museum, *Handicapped* (Washington, D.C.: USHMM Pamphlet Series, n.d.), 2.

10. Biesold surveyed more than 1,200 deaf persons who were sterilized. See Biesold, *Crying Hands.*

11. Quoted in Biesold, *Crying Hands,* 29.

12. The German world for deaf is *taub,* but hearing Germans commonly use the word *taubstumm,* which is translated as "deaf and dumb." German Deaf community historian Jochen Muhs informs me that contemporary German deaf people prefer the use of the term *Gehörlos.* This is comparable to the English *hearing loss* or *hearing impaired.*

13. However, the teams, like deaf organizations such as REGEDE, excluded Jewish deaf Germans.

14. For the complete article, see Kurt Lietz, "The Place of the School for the Deaf in the New Reich," trans. Tobias Brill, in Ryan and Schuchman, *Deaf People,* 114–20.

PART 3

THE SCIENCE
OF GENETICS
AND DEAFNESS

INTRODUCTION

Genetics and disability began to interact in new ways in the late-twentieth century, as the Human Genome Project identified all the genes humans carry and mapped the sequences of chemicals that determine genetic structures. Speculation about designer babies and discrimination based on genetic profiles became staples of popular culture. Some of these manifestations of the new science of genetics are discussed in the last two essays of this volume. The genetics of deafness in particular, however, took a sudden and remarkable turn in the 1990s with the discovery of a small gene called connexin 26, which is believed to be responsible for nearly half of all genetic, prelingual deafness today (and perhaps more in the future, as physician and geneticist Walter Nance speculates). The discovery of this gene, which is easily detected with a blood test and can be identified prenatally, raises ethical, counseling, and public policy questions.

Orit Dagan and Karen Avraham review the classification, causes, and frequency of the most common genetic causes of auditory abnormalities in the first essay in this section. They explain that scientific data related to human hearing have been developed primarily from the study of family histories or, more recently, from studies of mice, whose auditory structures and genetics are similar to those of humans. They caution that the relationship between environmental factors and genetic factors in hearing loss is not completely understood. The "genetic component" of conditions such as presbycusis and noise-induced hearing loss warrants much further study. Finally, they recognize that research on genetics and deafness raises ethical questions.

Walter Nance is the most influential American student of genetics and deafness. His work has influenced generations of scholars, and his essay in this section demonstrates why. He reviews the importance of connexin 26, discusses the most famous collection of genetic data on deaf people ever constructed—that of Edward Allen Fay at Gallaudet in the late-nineteenth century—and reevaluates Alexander Graham Bell to suggest how the epidemiology of genetic deafness may change through time.

Nance uses the Fay data, data from modern genetic studies, and computer simulations to conclude that the frequency of connexin 26 deafness has doubled in the United States over the past 100 years. This has resulted, he believes, from assortative mating, or the tendency of people to chose marriage partners with whom they can communicate easily. Until the late-twentieth century, geneticists commonly believed that deafness had so many possible genetic causes (hundreds) that assortative mating did not matter and that, therefore, Bell's warning about deaf-deaf marriages was, Nance writes, "unfounded." Now that the crucial importance of connexin 26 is understood, however, Nance concludes, Bell "may have been correct." Importantly, however, Nance does not attach normative value to his conclusion, for he believes that selective mating is not necessarily detrimental, and gives another historical example to emphasize his point.

THE COMPLEXITY OF HEARING LOSS FROM A GENETICS PERSPECTIVE

Orit Dagan and Karen B. Avraham

Molecular biology has provided tremendous tools for answering biological questions in the past twenty-five years. Today, we have had a revolution in this area with the Human Genome Project and the sequencing of the human genome—the identification of the alphabet that defines our genes—and hence every part of our body, including of course, the inner ear. Virtually all diseases and disorders have a genetic component. This is not to say that the environment does not contribute a large part; rather, it is often how the two interact that will determine the severity of a disease or disorder. For example, three types of hearing loss represent different proportions of the contribution of genetics versus environment: Pendred syndrome, which is associated with thyroid disorders and has a major genetic component due to mutations in the pendrin gene;[1] age-related hearing loss or presbycusis, for which both genetics and environment may play a near equal role; and hearing loss due to meningitis, which is largely environmental and caused by exposure to a virus. In addition, our susceptibility to hearing loss due to noise or other trauma may be due to genetic factors.

The genetic basis of the audio-vestibular system needs further study. Scientists would like to understand how the inner ear forms, from the time it is just a round "hole" called an otocyst, until it becomes an intricate six-part ear that works with exquisite precision to allow mammals to hear. The ear is one of the most complex organs of the body, and while we have learned a great deal about it in past years, we are still a long way from completely understanding how it functions.

A genome is the entire set of genetic material in a living creature. The cell contains a nucleus, which is essentially its "heart." Each nucleus contains the organism's entire genetic material, arranged on chromosomes that pass from generation to generation, allowing us to inherit genetic material from our parents. Each gene codes for a different protein. The proteins function to transport molecules (such as myosins) along the cytoskeleton in the cell, form bridges between cells (such as gap junctions) so that molecules can pass through, tell other genes (such as transcription factors) when to be turned on in a cell and when to be turned off, and many other functions. It is these types of genes that have mutations, changes in the DNA leading to dysfunction of the protein they encode, which can lead to hearing impairment.

The Human Genome Project, spanning a period of approximately fifteen years, had several goals: to identify all the genes in the human genome, to identify all the genes in human DNA, to determine the sequences of the three billion bases that make up human DNA, to understand gene structure and variation, to decipher comparative genomics (human vs. other species), and to address the ethical, legal, and social issues (ELSI) that may arise from the project (www.ornl.gov/hgmis/elsi/elsi.html).[2]

Many important discoveries were made as a consequence of determining the sequence of all the nucleotides—the chemicals GATC that make up the "alphabet" of the genome in different combinations of these four letters—of the human genome.[3] One is that we have only 30,000–40,000 genes, which is far fewer than was expected, especially since mice have the same number and worms have about half this amount. This discovery has raised many questions about what differentiates humans from these creatures, other than just the genes humans have. Other discoveries have shown that there is a higher mutation rate in males than in females, and many of the genes in the human and mouse genomes are similar; encoded proteins between mouse and human have a median amino acid identity of 78.5 percent.[4]

In the field of hearing research, scientists are interested in knowing how many genes are involved in the formation and growth of the inner ear and how these genes work together to create auditory ability. We are also interested in discovering the specific proteins the genes encode, those responsible for the formation of the inner ear and its proper functioning. Using mice as models for studying human hearing impairment has helped address these questions.[5]

GENETIC HEARING LOSS

There are many causes of hearing loss, both environmental and genetic. Ototoxic drugs, rubella during pregnancy, excessive noise, and meningitis are among the environmental causes. Hearing loss may also be due to a mutation in a gene, leading to an abnormal protein or the loss of a protein.

Approximately 60 percent of hearing loss is genetic. Genetic hearing loss is classified as either syndromic or nonsyndromic. Syndromic hearing loss includes other features, such as retinitis pigmentosa (for Usher syndrome), kidney abnormalities (for Alport syndrome), and pigment defects (such as Waardenburg syndrome). Approximately 400 syndromes have been described that include hearing loss (Online Mendelian Inheritance in Man: www.ncbi.nlm.nih.gov/omim/). Most genetic hearing loss, however, is nonsyndromic, where hearing loss might be associated with vestibular dysfunction, but no other signs.

Nonsyndromic hearing loss is very heterogeneous, that is, mutations in many genes lead to hearing impairment. This can be seen in different modes of inheritance (recessive, dominant, X-linked, or mitochondrial), different ages of onset (prelingual, postlingual), differences in severity (mild, moderate, severe, or profound), differences in stability, and differences in site affected (middle ear for conductive hearing loss or inner ear for sensorineural hearing loss).

There are two major strategies to identify genes associated with hearing loss: working with families with inherited deafness and working with mouse models. The latter is possible because the structure and function of the inner ear in humans and mice are very similar, and most of the genes found in humans have a partner in the mouse. In fact, there are hundreds of mouse models for human hearing loss.[6]

The chromosomal locations for over ninety loci associated with human hearing loss are known to date; they are randomly distributed throughout the human genome. DFN stands for "deafness" (and X-linked deafness), DFNA for autosomal dominant inheritance, and DFNB for autosomal recessive inheritance. Details of the loci are updated regularly in the Hereditary Hearing Loss Homepage (http://www.uia.ac.be/dnalab/hhh/). The loci are numbered in the order in which they were discovered: *DFNA1* was mapped in 1992, and *DFNA51* was recently mapped in our laboratory (Dagan and Avraham, unpublished data).[7] To date, thirty-three genes have been cloned for both autosomal recessive and dominant hearing loss (figure 1).

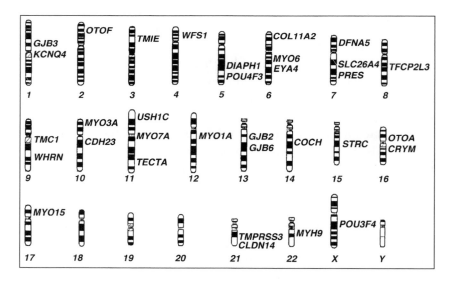

Figure 1. Genes associated with hearing loss in the human population. The genes are randomly distributed throughout the twenty-two autosomes and the X chromosome. Mutations in thirty-three genes are associated with different forms of nonsyndromic hearing loss. These mutations are inherited in an autosomal dominant or recessive manner, or through chromosome X. The names of each gene listed next to the chromosome it lies on are the gene symbols. For the full name of each gene, the protein each gene encodes, and further information, see the Hereditary Hearing Loss Homepage.

What is particularly useful for researchers is that there are mouse models that represent abnormalities in different parts of the ear, so that we have the ability to learn about many aspects of the ear. For example, there are mice with middle ear defects (conductive hearing loss), mice that lose their hair cells after birth and are a model for progressive hearing loss (such as the Beethoven mouse), and mice that have vestibular dysfunction due to elongation of their stereocilia or because of malformations in the bony structure of the vestibule.[8]

AUTOSOMAL RECESSIVE HEARING LOSS

As discussed earlier, hearing loss can be inherited in an autosomal dominant or recessive fashion. To date, twenty genes are known to cause autosomal recessive hearing loss. Most forms of this hearing loss are prelingual, although exceptions do exist (for example, *DFNB30* is associated with onset of hearing loss in adulthood and is progressive).[9] However, one of the most significant discoveries in the genetics of deafness is that of connexin 26, since mutations in this gene are found in 30–50 percent of children with prelingual hearing loss.[10] For cases where there is a family history, most deafness is inherited in an autosomal recessive mode. Connexin 26 mutations lead to hearing loss that ranges from severe to profound, is usually stable in nature, and is not associated with morphological abnormalities in the ear.[11] Today, this is usually the only gene checked for diagnostically. Connexin 26 is a very small gene that is easy to check for mutations.[12] If there are no mutations in this gene, the search for the mutation causing hearing loss is more difficult, since there is no one prevalent gene other than connexin 26.

Connexin 26, though present in all cells of the body, forms an essential component in the inner ear, since mutations lead to hearing loss. Six connexins make up a connexon, which is present on the surface of cells. In order for ions and other molecules to pass from one cell to the next, a connexon from one cell must connect with a connexon from another cell. There are many connexins in our body, in almost all cell types. The inner ear has several connexins, and indeed mutations have been found in two others, connexin 30 and connexin 31, although these are quite rare.[13] The discovery of connexin 26 in deafness has changed genetic counseling, since we now have the ability to easily check for mutations in this gene. This has raised ethical questions, some of which will be discussed later.

AUTOSOMAL DOMINANT INHERITANCE

Hearing loss can also be inherited in an autosomal dominant fashion. To date, seventeen genes are known to cause autosomal dominant hearing loss. This form of hearing loss is usually postlingual and progressive. There does not appear to be any one prevalent gene for autosomal dominant hearing loss, and a particular mutation is often found in only one extended family. For example, a *POU4F3* mutation has been found in one extended family, suggesting the presence of a founder mutation.[14] Family H is a large Israeli family with progressive hearing loss. The onset is at around age eighteen, and it progresses to severe hearing loss by the age of fifty.[15] This family participated in the research project because they were interested in knowing what gene is involved in their hearing loss.

Our laboratory, in collaboration with scientists from the University of Washington in Seattle and the National Institutes of Health, discovered an 8-bp deletion in the *POU4F3* gene. This gene is a transcription factor, whose function is to regulate the expression of other genes in the cell. Recently we created a cell model in order to begin to understand the mechanism of the mutation.[16] How does the 8-bp deletion in *POU4F3* lead to hearing loss in this family? The mutant protein, instead of being localized in the nucleus of the cell, is in the cytoplasm of the cell. If this transcription factor is required to regulate other genes in the cells of the ear, the loss of appropriate amounts of *POU4F3* may compromise this function. A mouse model for *POU4F3*-associated deafness was constructed by removing this gene from its genome (gene-targeted mutagenesis).[17] The mice are born with degenerated hair cells, which eventually die due to premature cell death (apoptosis). Therefore the cell and mouse models led to the conclusion that a mutation in *POU4F3* leads to hair cell death, and that this is the cause of the hearing loss in Family H.

Another example of a gene involved in hearing loss is myosin VI, a gene that encodes a molecular motor that is responsible for transporting molecules within the cell. In the hair cell, the myosins are responsible for transporting necessary molecules to the stereocilia, the tiny projections that come out of the top of the hair cell that move in

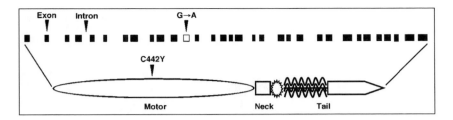

Figure 2. The molecular motor myosin VI is involved in hearing loss. The gene is composed of thirty-three coding exons (top) that encode a protein. This protein is comprised of a motor domain, a neck domain, and a tail. The motor domain allows this protein to bind to actin, and the unique tail region confers its specificity. A G to A transition was found in exon twelve of the myosin VI gene in an Italian family with progressive autosomal dominant HL. This missense mutation causes a cysteine (C) to tyrosine (Y) amino acid change at position 442, which may affect the stability of the myosin VI protein.[19]

response to sound. The role of myosin VI in hearing loss was first discovered in 1995, when it was found to be missing in the deaf Snell's waltzer mouse.[18] A few years later, an Italian family was identified with a mutation in myosin VI, with postlingual dominant hearing loss (figure 2).[19] Recently, a number of Pakistani families with inherited recessive hearing loss were found to have myosin VI mutations as well.[20] However, the mouse has provided the most clues about what actually happens to the hair cells due to a myosin VI mutation. The stereocilia of Snell's waltzer mice fuse, beginning one day after birth, which presumably does not allow the channels between the tops of the stereocilia to function properly (figure 3).[21] We predict that the same phenomenon occurs in humans, leading to hearing impairment.

ETHICAL QUESTIONS

The discovery of genes associated with hearing loss has raised ethical questions regarding the use of diagnostics to predict the onset of hearing loss. First, are individuals with hearing loss interested in such testing? Is the interest in genetic counseling and testing different among Deaf adults compared with deaf children of hearing parents? These and other questions were raised in a study performed in the United Kingdom.[22] In the Israeli population, a study has been carried out on the interest and motivations of hearing parents of deaf children to choose genetic testing

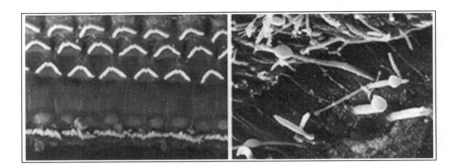

Figure 3. The structure of cochlear hair cells is affected due to a myosin VI mutation. Scanning electron microscopy allows for visualization of the morphology of the surface of hair cells. The left panel shows a group of hair cells from a hearing mouse, with three rows of outer hair cells and one row of inner hair cells. The right panel shows a group of hair cells from a deaf Snell's waltzer mouse with mutant myosin VI at the age of twenty days. Cells are disorganized, and the fusion of stereocilia reveals giant stereocilia. (Adapted with permission from Self et al. 1999.)

and prenatal diagnosis.[23] There was a very high interest (87 percent) in genetic testing for deafness among Israeli Jewish parents. The reasons for undertaking a test varied significantly between nonreligious and religious individuals, however.

Should carrier screening be carried out in populations with high carrier rates? This issue was raised in Israel, as there is a prevalent mutation, 167delT in the connexin 26 gene in the Jewish Ashkenazi population.[24] A decision was made not to perform screening on a national level, although some genetic clinics do offer screening for the mutation. Should children from families with inherited progressive hearing loss be tested for the mutation as children? As with other late-onset diseases and disorders, testing is not being performed in children, as there is no intervention that would delay the onset of hearing loss. Furthermore, testing children would infringe on their personal autonomy and on the privacy of their own genetic information, since their parents would be making the decision for them. These issues have been raised for late-onset genetic diseases, such as Huntington's disease.[25] And finally, will this work lead to an attempt to treat hearing loss? The answer to this question is unequivocally yes, as there have already been a number of studies attempting to use knowledge of the genes involved in hearing loss for therapeutics.[26] What must be emphasized is that this approach will be used only for those who are interested in such therapy, such as individuals with late-onset hearing loss and the ageing population. Furthermore, research may also provide improvements in the efficacy of the cochlear implant, as new methods are being developed to cause neurons to grow towards cochlear implants.[27]

THE FUTURE FOR RESEARCH IN THE HEARING FIELD

What does the future hold for research in the field of hearing loss? Due to the complexity of the inner ear, there is still a great deal of information required in order to understand it fully. Of the ninety loci mapped for monogenic (one gene only) forms of hearing loss, fifty-three of the genes still need to be cloned. There are other forms of hearing loss that are believed to have a genetic component, such as otosclerosis, which leads to conductive hearing loss and, at a late stage, to sensorineural hearing loss.[28] Ménierè's disease, hearing loss associated with vestibular attacks, may have a genetic basis, although this has been more difficult to resolve. Perhaps the greatest impact research will have will be on presbycusis (age-related hearing loss) and noise-induced hearing loss. As the population ages, more people will be affected by these conditions, and the new genomic techniques now available may address the causes of hearing loss in this group.[29]

CONCLUSIONS

Research into hereditary hearing loss has expanded dramatically in the past decade—the first autosomal locus was mapped in 1992, and today, over ninety loci have been mapped. In almost half of these cases, the gene with mutations leading to hearing loss has been identified, and we have acquired knowledge about the function of the protein that the gene encodes. As knowledge of the inner ear and its normal function grows, so does awareness of the complexity of the many different cell types and multiple proteins acting in concert. Many new techniques in molecular and cell biology, including microarrays and protein chips, will allow the study of the networks of protein interactions. This will help develop a thorough understanding of the mechanisms of the inner ear and how genetic mutations lead to hearing impairment.

ACKNOWLEDGMENTS

We would like to thank the members of the Avraham laboratory who have contributed over the years to the work described in this review and to colleagues who have helped us identify genes: Karen Steel, Mary-Claire King, Thomas Friedman, Moien Kanaan, and Paolo Gasparini. Research in the Avraham laboratory is supported by the National Institutes of Health (R01 DC005641), the Israel Science Foundation, the Israel Ministry of Science, and the National Organization for Hearing Research.

REFERENCES

1. Everett LA, Glaser B, Beck JC, Idol JR, Buchs A, Heyman M, Adawi F, Hazani E, Nassir E, Baxevanis AD, Sheffield VC, and Green ED. Pendred syndrome is caused by mutations in a putative sulphate transporter gene (PDS). Nature Genetics. 1997;17:411–422.

2. Ruddle F. Hundred-year search for the human genome. Annual Review of Genomics and Human Genetics. 2001;2:1–8.

3. Lander ES, Linton LM, Birren B, Nusbaum C, Zody MC, Baldwin J, Devon K, Dewar K, Doyle M, FitzHugh W, Funke R, Gage D, et al. Initial sequencing and analysis of the human genome. Nature. 2001;409:860–921.

4. Waterston RH, Lindblad-Toh K, Birney E, Rogers J, Abril JF, Agarwal P, Agarwala R, Ainscough R, Alexandersson M, An P, Antonarakis SE, Attwood J, et al. Initial sequencing and comparative analysis of the mouse genome. Nature. 2002;420:520–562.

5. Avraham KB. Mouse models for deafness: Lessons for the human inner ear and hearing loss. Ear and Hearing. 2003;24:332–341.

6. Kiernan AE and Steel KP. Mouse homologues for human deafness. Adv Otorhinolaryngol. 2000;56:233-243; Ahituv N and Avraham KB. Mouse models for human deafness: current tools for new fashions. Trends in Molecular Medicine. 2002;8:447–451.

7. Leon PE, Raventos H, Lynch E, Morrow J, and King MC. The gene for an inherited form of deafness maps to chromosome 5q31. Proceedings of the National Academy of Sciences of the United States of America. 1992;89:5181–5184.

8. Vreugde S, Erven A, Kros CJ, Marcotti W, Fuchs H, Kurima K, Wilcox ER, Friedman TB, Griffith AJ, Balling R, Hrabe De Angelis M,

Avraham KB, and Steel KP. Beethoven, a mouse model for dominant, progressive hearing loss DFNA36. Nature Genetics. 2002;30:257–258; Kiernan AE, Ahituv N, Fuchs H, Balling R, Avraham KB, Steel KP, and Hrabe De Angelis M. The Notch ligand Jagged1 is required for inner ear sensory development. Proceedings of the National Academy of Sciences of the United States of America. 2001;98:3873–3878.

9. Walsh T, Walsh V, Vreugde S, Hertzano R, Shahin H, Haika S, Lee MK, Kanaan M, King MC, and Avraham KB. From flies' eyes to our ears: Mutations in a human class III myosin cause progressive nonsyndromic hearing loss DFNB30. Proceedings of the National Academy of Sciences of the United States of America. 2002;99:7518–7523.

10. Denoyelle F, Weil D, Maw MA, Wilcox SA, Lench NJ, Allen-Powell DR, Osborn AH, Dahl HH, Middleton A, Houseman MJ, Dode C, Marlin S, Boulila-ElGaied A, Grati M, Ayadi H, BenArab S, Bitoun P, Lina-Granade G, Godet J, Mustapha M, Loiselet J, El-Zir E, Aubois A, Joannard A, Petit C et al. Prelingual deafness: High prevalence of a 30delG mutation in the connexin 26 gene. Human Molecular Genetics. 1997;6:2173–2177; Estivill X, Fortina P, Surrey S, Rabionet R, Melchionda S, D'Agruma L, Mansfield E, Rappaport E, Govea N, Mila M, Zelante L, and Gasparini P. Connexin-26 mutations in sporadic and inherited sensorineural deafness. Lancet. 1998;351:394–398.

11. Denoyelle F, Lina-Granade G, Plauchu H, Bruzzone R, Chaib H, Levi-Acobas F, Weil D, and Petit C. Connexin 26 gene linked to a dominant deafness. Nature. 1998;393:319–320; Cohn ES and Kelley PM. Clinical phenotype and mutations in connexin 26 (DFNB1/GJB2), the most common cause of childhood hearing loss. American Journal of Medical Genetics. 1999;89:130–136.

12. Sobe T, Vreugde S, Shahin H, Berlin M, Davis N, Kanaan M, Yaron Y, Orr-Urtreger A, Frydman M, Shohat M, and Avraham KB. The prevalence and expression of inherited connexin 26 mutations associated with nonsyndromic hearing loss in the Israeli population. Human Genetics. 2000;106:50–57.

13. del Castillo I, Villamar M, Moreno-Pelayo MA, del Castillo FJ, Alvarez A, Telleria D, Menendez I, and Moreno F. A deletion involving the connexin 30 gene in nonsyndromic hearing impairment. New England Journal of Medicine. 2002;346:243–249; Xia JH, Liu CY, Tang BS, Pan Q, Huang L, Dai HP, Zhang BR, Xie W, Hu DX, Zheng D, Shi XL, Wang DA, Xia K,

Yu KP, Liao XD, Feng Y, Yang YF, Xiao JY, Xie DH, and Huang JZ. Mutations in the gene encoding gap junction protein beta-3 associated with autosomal dominant hearing impairment. Nature Genetics. 1998;20:370–373.

14. Vahava O, Morell R, Lynch ED, Weiss S, Kagan ME, Ahituv N, Morrow JE, Lee MK, Skvorak AB, Morton CC, Blumenfeld A, Frydman M, Friedman TB, King MC, and Avraham KB. Mutation in transcription factor POU4F3 associated with inherited progressive hearing loss in humans. Science. 1998;279:1950–1954.

15. Frydman M, Vreugde S, Nageris BI, Weiss S, Vahava O, and Avraham KB. Clinical characterization of genetic hearing loss caused by a mutation in the POU4F3 transcription factor. Archives of Otolaryngology—Head and Neck Surgery. 2000;126:633–637.

16. Weiss S, Gottfried I, Mayrose I, Khare SL, Xiang M, Dawson SJ, and Avraham KB. The DFNA15 deafness mutation affects POU4F3 protein stability, localization and transcriptional activity. Molecular and Cellular Biology. 2003;23:7957-7964.

17. Erkman L, McEvilly RJ, Luo L, Ryan AK, Hooshmand F, O'Connell SM, Keithley EM, Rapaport DH, Ryan AF, and Rosenfeld MG. Role of transcription factors Brn-3.1 and Brn-3.2 in auditory and visual system development. Nature. 1996;381:603–606; Xiang M, Gao W-Q, Hasson T, and Shin JJ. Requirement for Brn-3c in maturation and survival, but not in fate determination of inner ear hair cells. Development. 1998;125:3935–3946.

18. Avraham KB, Hasson T, Steel KP, Kingsley DM, Russell LB, Mooseker MS, Copeland NG, and Jenkins NA. The mouse Snell's waltzer deafness gene encodes an unconventional myosin required for structural integrity of inner ear hair cells. Nature Genetics. 1995;11:369–375.

19. Melchionda S, Ahituv N, Bisceglia L, Sobe T, Glaser F, Rabionet R, Arbones ML, Notarangelo A, Di Iorio E, Carella M, Zelante L, Estivill X, Avraham KB, and Gasparini P. MYO6, the human homologue of the gene responsible for deafness in Snell's waltzer mice, is mutated in autosomal dominant nonsyndromic hearing loss. American Journal of Human Genetics. 2001;69:635–640.

20. Ahmed ZM, Morell RJ, Riazuddin S, Gropman A, Shaukat S, Ahmad MM, Mohiddin SA, Fananapazir L, Caruso RC, Husnain T, Khan SN, Griffith AJ, Friedman TB, and Wilcox ER. Mutations of MYO6 are associated with recessive deafness, DFNB37. American Journal of Human Genetics. 2003;72:1315–1322.

21. Self T, Sobe T, Copeland NG, Jenkins NA, Avraham KB, and Steel KP. Role of myosin VI in the differentiation of cochlear hair cells. Developmental Biology. 1999;214:331–341.

22. Middleton A, Hewison J, and Mueller RF. Attitudes of deaf adults toward genetic testing for hereditary deafness. American Journal of Human Genetics. 1998;63:1175–1180.

23. Dagan O, Hochner H, Levi H, Raas-Rothschild A, and Sagi M. Genetic testing for hearing loss: Different motivations for the same outcome. American Journal of Medical Genetics. 2002;113:137–143.

24. Sobe T, Erlich P, Berry A, Korostichevsky M, Vreugde S, Avraham KB, Bonne-Tamir B, and Shohat M. High frequency of the deafness-associated 167delT mutation in the connexin 26 (GJB2) gene in Israeli Ashkenazim. American Journal of Medical Genetics. 1999;86:499–500.

25. Bloch M and Hayden MR. Opinion: Predictive testing for Huntington's disease in childhood: Challenges and implications. American Journal of Human Genetics. 1990;46:1–4.

26. Kawamoto K, Kanzaki S, Yagi M, Stover T, Prieskorn DM, Dolan DF, Miller JM, and Raphael Y. Gene-based therapy for inner ear disease. Noise Health. 2001;3:37–47; Shou J, Zheng JL, and Gao WQ. Robust generation of new hair cells in the mature mammalian inner ear by adenoviral expression of Hath1. Molecular and Cellular Neuroscience. 2003;23:169–179.

27. Kawamoto K, Ishimoto S, Minoda R, Brough DE, and Raphael Y. Math1 gene transfer generates new cochlear hair cells in mature guinea pigs in vivo. Journal of Neuroscience. 2003;23:4395–4400.

28. Van Den Bogaert K, Govaerts PJ, De Leenheer EM, Schatteman I, Verstreken M, Chen W, Declau F, Cremers CW, Van De Heyning PH, Offeciers FE, Somers T, Smith RJ, and Van Camp G. Otosclerosis: A genetically heterogeneous disease involving at least three different genes. Bone. 2002;30:624–630.

29. DeStefano AL, Gates GA, Heard-Costa N, Myers RH, and Baldwin CT. Genomewide linkage analysis to presbycusis in the Framingham Heart Study. Archives of Otolaryngology—Head and Neck Surgery. 2003;129:285–289.

THE EPIDEMIOLOGY OF HEREDITARY DEAFNESS
The Impact of Connexin 26 on the Size and Structure of the Deaf Community

Walter E. Nance

This paper will discuss briefly what we now know about the causes of deafness and how they can change with time. It will not dwell upon the many different forms of genetic deafness that have been identified as part of the Human Genome Project, but it will review what is known about the commonest form of genetic deafness, connexin 26, and share some ideas about why that form of deafness is so frequent. In the process, we will see what an important role genetic studies at Gallaudet University have played in understanding the genetic epidemiology of deafness. Finally, the paper may contribute a new perspective on the ambiguous role that Alexander Graham Bell played in the history of deaf culture.

Edward Allen Fay

ETIOLOGIC HETEROGENEITY

Deafness has many recognized causes, and an important goal of genetic analysis has been to attempt to distinguish causes that are largely genetic from those that are environmental in origin. Long before we learned to map and clone human genes effectively, it was possible to at least estimate the overall contribution of genetic factors to deafness with a high degree of accuracy through the systematic collection and analysis of data on the frequency and distribution of deafness within families. Without question, the largest and most valuable data set of this type that has ever been assembled are the 5,000 pedigrees collected by Edward Allen Fay, a professor at Gallaudet College, and published in a book titled *Marriages of the Deaf in America* in 1898.[1] These data were collected with the active support of Alexander Graham Bell and were analyzed for the first time before the rediscovery of Gregor Mendel's work in 1900. As a consequence, Fay's insightful interpretations did not benefit from knowledge of the then-new theory of Mendelian Inheritance. However, the questionnaires that he used were so well designed that the data have been reanalyzed repeatedly during the past 100 years using progressively more sophisticated methods.

In the contemporary approach to the analysis of family data of the type Fay collected, we assume that the deafness is genetic in all families with two or more deaf siblings. Then, using knowledge about how genetic traits are transmitted in families, we estimate what proportion of the families with only one deaf child are also genetic cases in which by chance only one child is deaf. This method is analogous to using knowledge about the density of ice and water to estimate the total size of an iceberg by measuring the size of the part that is above the water. When analyses of this type are performed, we can estimate that about half of all cases of profound deafness in this country during the nineteenth century were genetic in origin.[2] Of course we know that some of the environmental causes of deafness, such as cytomegalovirus (CMV) infections and the congenital rubella syndrome, can be epidemic in nature, so that the proportions of genetic and environmental deafness can vary tremendously from time to time. Thus, as you might expect, the proportion of genetic cases was much lower during the last rubella epidemic.

Genetic deafness may be dominant, recessive, X-linked, or carried through mitochondrial inheritance, and it may be either syndromic or nonsyndromic.[3] Further analyses of Fay's data alone permit at least crude estimates of the proportions of the former groups, if not the latter, types of genetic deafness. More recently, in the mid-1960s, a demographer named Augustine Gentile at Gallaudet initiated the systematic annual collection of demographic data on the 40,000–50,000 children in educational programs for the deaf throughout the nation. During the past forty years, the subsequent directors of the Annual Survey of Deaf and Hard of Hearing Children and Youth have been very receptive to the collection of information on deafness in the families of these children. On two occasions, data on more than 12,000 families have been collected, resulting in the largest samples of family data on a single genetic trait that have ever been subjected to genetic analysis.[4] Currently, with support from an NIH grant and in collaboration with scientists at Gallaudet, we are also using families identified through the survey to establish a nationwide repository of DNA samples from deaf subjects and their families to search for new deafness genes. Finally, it is important to emphasize that deafness is not always either genetic or environmental, and a growing number of important examples are being recognized in which specific genes may predispose an individual to become deaf if they are exposed to specific environmental factors.[5]

Augustine "Gus" Gentile

CONNEXIN DEAFNESS

In view of the large number of genes that can cause deafness, the discovery that genetic changes involving a single gene called connexin 26 are the cause of more than half of all cases of genetic deafness in many countries came as a surprise. The connexins are a family of related genes that code for the subunits of proteins that are required to form submicroscopic channels, called gap junctions, which allow ions and small molecules to flow freely between adjacent cells. At least four members of the connexin family are produced in the cochlea, where they play a critical roll in the perception of sound.[6] When data on the high frequency of connexin deafness first appeared, we decided to use the Fay data to confirm the reported findings in the United States in the nineteenth century. You might wonder how we can use data about people who have been dead for more than 100 years to estimate the frequency of a newly identified gene that can only be detected by DNA tests. It turns out that we have known for a long time that most deaf couples, nearly 90 percent, have hearing children. This can happen even if both parents have recessive deafness caused by two independent genes. About 7–10 percent of deaf couples have deaf and hearing children, while 4.2 percent have all deaf children and are, in fact, biologically incapable of having hearing children. This latter situation arises when both parents carry genes for the same form of recessive deafness and can only transmit copies of the same recessive gene to their children. If a gene has a frequency of p in the population, the chance of carrying two copies of the same gene would be p × p, or p^2. And the frequency of marriages between individuals with the same form of recessive deafness would then be $p^2 × p^2$, or p^4. Now, we had previously estimated the frequency of deaf-deaf marriages that could only have deaf offspring from the Fay data. So, if we make the extreme assumption that every single one of those marriages are between individuals with connexin deafness, we can obtain an estimate of the upper limit of the frequency of connexin deafness in the United States in the nineteenth century simply by taking the square root of 4.2 percent or about 17–19 percent. This figure is about half the reported current frequency based on actual testing of deaf subjects, suggesting that the frequency of connexin deafness may have increased in the United States during the past 200 years.[7]

According to Darwin's theory of evolution, the frequency of a trait in a population will depend on the balance between the occurrence of new genetic changes or mutations for the trait and the "genetic fitness" of those who carry the trait. It seems likely that if you go back 1,000 years, the chance that an individual with profound deafness would grow up, marry, and produce offspring would have been very low. However, about 400 years ago, shared sign languages began to appear, schools for deaf people were established, and as a consequence, the social and economic circumstances of deaf people began to improve dramatically, along with their fertility.

For decades, geneticists have taught students that they should not be concerned that finding cures for serious genetic diseases, such as phenylketonuria or cystic fibrosis, will increase the frequency of those diseases in the population, because hundreds of generations would be required for the "relaxed selection" to have an appreciable effect on the population frequency. However, this formulation of the problem assumes that the population will be mating at random. In the case of deafness, though, the introduction of sign language was accompanied by the onset of intense linguistic homogamy, or the tendency of deaf persons to select marriage partners who are also fluent in sign language.

We have conducted computer simulations that show that the combination of improved fitness and linguistic homogamy can very greatly accelerate the slow increase in the gene frequency that would occur if mating were to occur at random (figure 1).[8] Our findings indicate that

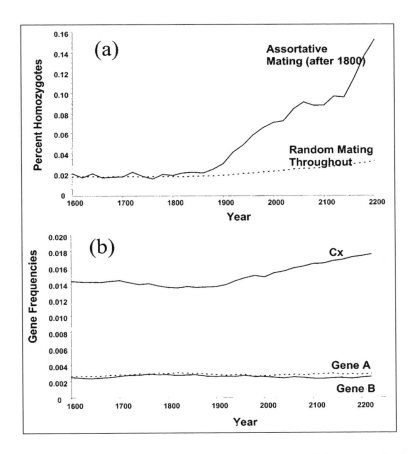

Figure 1. Computer simulations showing: (a) The expected changes in the frequency of a specific form of recessive deafness after improvement in the fertility of deaf people, assumed in the simulation to have begun in 1800, in the presence and absence of intense linguistic homogamy (i.e., "assortative mating"); (b) Differences in the predicted effects on the frequency of a recessive gene that was initially the most common cause of recessive deafness and two other genes for less frequent forms of recessive deafness. Published with permission from the *American Journal of Human Genetics*. Originally published in W. E. Nance and M. J. Kearsey. Relevance of connexin deafness (DFNB1) to human evolution. *American Journal of Human Genetics*. 2004;74: 1081–1087.

this mechanism could, in fact, have doubled the frequency of connexin deafness in the past 200 years. It is important to emphasize that this mechanism will selectively increase the frequency of the commonest form of deafness in a population, but it could equally affect some other form of recessive deafness in a different population.

A very interesting example has been described in the village of Bengkala on the island of Bali, where 2 percent of the population of 2,385 villagers have a form of recessive deafness known as *DFNB3*, and 13 percent of the hearing population are carriers of the recessive gene. A remarkable feature of this population is that a long time ago, the inhabitants of Bengkala developed an indigenous sign language, which is now learned by both deaf and hearing villagers. As a result of their integration into the community, the fitness of deaf individuals is unimpaired, and deaf-deaf matings are much more frequent than chance would predict. With one or two interesting exceptions, all of the deaf-deaf marriages in this village produce all-deaf offspring, as you would expect if *DFNB3* were the only cause of deafness in this community.[9] Although random changes in gene frequency ("gene drift") probably played a roll in the initial survival of this gene, it is hard to escape the conclusion that once the recessive deafness appeared in Bengkala, and especially after sign language developed, the combination of improved fitness and assortative mating must also have contributed to the dramatic increase in the frequency of the gene and phenotype. Bengkala shows that this novel genetic mechanism could potentially increase the frequency of any form of recessive deafness, not just connexin deafness.

RELEVANCE TO HUMAN EVOLUTION

The apparent relevance of the mating structure of the deaf community to the frequency of connexin deafness made us wonder whether there are other examples where improved fitness and assortative mating have had an influence on human genes. In the past two years, several genes that are important for the development of speech, including one called *FOXP2*, have been identified. Evolutionary biologists now agree that the acquisition of speech was the most important single event that led to the evolutionary divergence of Homo sapiens from other higher primates. But the mechanisms that have allowed speech and related attributes of the human brain to evolve with such extraordinary speed during the past 50,000–100,000 years have always been a profound mystery. The answer,

Alexander Graham Bell

of course, now seems clear. If you were the first primate to carry a gene for speech, wouldn't you have wanted to select a partner with whom you could communicate? We think this is exactly what happened in the case of the *FOXP2* gene on chromosome 8, and molecular studies have now shown that the protein produced by the *FOXP2* gene in humans differs from all higher primates in two amino acid subunits, and confirmed the fact that that the speed with which this evolutionary change has taken place exceeds that of all other genes studied to date. The *FOXP2* gene has been evolving more rapidly than any other gene at a rate that is more than 100 times the average.[10] We thus believe that selective mating is not just a curious anomaly of Deaf culture but is actually an essential feature of human evolution.

THE AMBIGUOUS ROLE OF ALEXANDER GRAHAM BELL

These new insights force us to reevaluate the ambiguous role that Bell played in the history of deafness. Bell devoted most of his professional career to the education of deaf children. Some have said that he was motivated to invent the telephone by his desire to correct the hearing loss of his mother. In 1876, long before the rediscovery of Mendel's work, Bell wrote an article in the *Proceedings of the National Academy of Science* in which he expressed his concern that intermarriages among deaf people might someday create a deaf variety of the human race.[11] He considered, but rejected, the possibility of prohibiting marriages among deaf people, but did raise the possibility of educating deaf children in regular schools. This was a revolutionary idea at the time, but it no longer seems so extreme when it is referred to as "mainstreaming." Bell also consorted with scientists who were identified with the eugenics movement, and undoubtedly made many statements that appear politically incorrect when read now. Most geneticists during the past century have considered his concerns about the frequency of deafness to be unfounded because of the very large number of genes involved in deafness. As noted previously, it now appears that he may have been correct, at least with respect to the commonest form of deafness. Even if he was wrong, it seems absurd to brand a person as a eugenicist who was advocating more random mating of the deaf and hearing populations. This is the exact opposite of Nazi proposals to create a master race by selective breeding. But evidently, those who enjoy going back into history and making moral judgments using contemporary standards are not concerned about such inconsistencies.

ETHICAL ISSUES

If these ideas are correct, they mean that if the deaf community so chose, it could create a deaf variety of the human race. For example, if 100 like-minded couples with connexin deafness immigrated to an unpopulated island, they could create a deaf variety of the human race in a single generation. I am not advocating this, nor do I think it will happen, but it is a dramatic example of how we can all be empowered by genetic knowledge in ways that were never before possible. It is reasonable to ask whether this empowerment is associated with any responsibilities. For example, those of us who can hear might well ask ourselves what attitude a society should take towards a pattern of marriages that has doubled the frequency of the commonest form of genetic deafness in the past 200 years. Perhaps my answer to this ethical dilemma may surprise you. We know very well that the tendency of Jews to marry one another has kept the frequency of Tay Sachs disease at a much higher level than it would be if marriages occurred at random. Similarly, the frequency of sickle-cell anemia is much higher in our country than it would be if African Americans married at random. Unless society is prepared to outlaw racial and ethnic homogamy, therefore, I can see no rational basis for proscribing marriages among deaf people. Even if it has increased the frequency of connexin deafness, this is simply a price we must pay for the freedom of association, which is a right upon which our culture places a very high value.

Modern critics of eugenics frequently complain that the movement was based on false or at least incomplete genetic knowledge. If I am right about the essential role that selective marriages for speech played in recent human evolution, these critics may now have to reexamine the scientific basis of the value system that they themselves are promoting. If assortative mating for speech did occur in our evolutionary past, it clearly was a natural phenomenon that was conducted without any knowledge whatsoever about its genetic basis or the profound eugenic consequences that would result. I want to make it clear that I am not for a moment condoning the tragic consequences of the perversion of eugenics during the twentieth century. But to conclude that all possible examples of selective mating are necessarily evil would incriminate a phenomenon that may have been one of the most important forces in human evolution. My own view about many of the seemingly intractable ethical issues that genetics has posed is that as long as decisions about issues such as marital choices or whether or not to use specific genetic technologies are made by competent, fully informed individuals, I am willing to live with the consequences. To the extent that those decisions are made for all of us by presidents, politicians, theologians, ethicists, geneticists, physicians, or other experts, I think there is great room for mischief. So the answer, in my view, is more conferences where divergent views can be expressed and each individual can make up his or her own mind.

RARE BOOKS COLLECTION

Finally, I want to tell you about a very exciting experience that I recently had. About thirty years ago, I was able to obtain from George Fellendorf at the Volta Bureau copies of the original questionnaires that Fay collected. For many years, we used these data for genetic analyses, and in some cases, have been able to track the family histories of deaf students that we see in the genetics clinic at Gallaudet University back to individuals contained in the Fay survey. Several years ago, I turned these records over to my colleague Kathy Arnos and was recently thrilled to learn that she, in turn, has given them to the rare books collection at Gallaudet, where they can be used by deaf students and alumni interested in exploring their family roots. Just last month, I saw the first student in our clinic use this resource to identify deaf ancestors who lived more than 100 years ago. If we are able to identify the specific form of genetic deafness that this student has, we may be able to guess which of her ancestors carried those genes. Genetics has always been exciting to me because, more than any other specialty in medicine, it allows you to understand the past and predict the future. The use of these records could not be a better example of this phenomenon.

REFERENCES

1. Fay EA. Marriages Among the Deaf in America. Washington: Gibson Brothers; 1898.

2. Rose SP. Genetic Studies of Profound Prelingual Deafness, (Ph.D. dissertation). Indianapolis (IN): Indiana University; 1975.

3. Nance WE. The genetics of deafness. Mental Retardation and Developmental Disabilities Research Reviews. 2003;9:109–19.

4. Marazita ML, Ploughman LM, Rawlings B, Remington E, Arnos KS, Nance WE. Genetic epidemiological studies of early-onset deafness in the U.S. school-age population. American Journal of Medical Genetics. 1993;46:486–91.

5. Pandya A, Radnaabazar J, Batsuri J, Dangaasuren B, Fischel-Ghodsian N, Nance WE. Mutation in the mitochondrial 12S rRNA gene in two families from Mongolia with matrilineal aminoglycosoide ototoxicity. Journal of Medical Genetics. 1997;34:169–72.

6. Pandya A, Arnos KS, Xia XJ, Welch KO, Blanton SH, Friedman TB, et al. Frequency and distribution of GJB2 (connexin 26) and GJB6 (connexin 30) mutations in a large North American repository of deaf probands. Genetics in Medicine. 2003;5:295–303.

7. Nance WE, Liu XZ, Pandya A. Relation between choice of a partner and high frequency of connexin deafness. Lancet;356:500–501.

8. Nance WE, Kearsey, MJ. Relevance of connexin deafness (DFNB1) to human evolution. American Journal of Human Genetics. 2004;74: 1081–1087.

9. Winata S, Arhya IN, Moeljopawiro S, Hinnant JT, Liang Y, Friedman TB, et al. Congenital non-syndromic autosomal recessive deafness in Bengkala, an isolated Balinese population. Journal of Medical Genetics. 1995;32:336–43.

10. Zhang J, Webb DM, Podlaha O. Accelerated protein evolution and the origins of human-specific features: Foxp2 as an example. Genetics. 2002;162:1825–35.

11. Bell AG. Upon the formation of a deaf variety of the human race. National Academy of Sciences Memoirs. 1883;2:177–262.

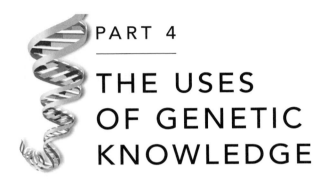

PART 4

THE USES
OF GENETIC
KNOWLEDGE

INTRODUCTION

The essays in this section demonstrate the intersection of genetic science and the lives of people who are labeled deaf or disabled, and they do so from widely divergent disciplinary and personal perspectives. The authors include scientists, genetics counselors, an anthropologist, and a science writer, Mark Willis, whose approach and conclusions are stark and personal. Lacking the distance from his subject—the "objectivity"—that the other authors in this section have, Willis states that if new studies of genetics are used to identify an "emerging genetic underclass," as another writer has claimed, then he could qualify for "class president or class clown" because of his genetic characteristics, his disabilities.

Kathleen Arnos and Arti Pandya offer a summary of genetic science related to deafness and then develop the argument for the genetic testing, evaluation, and counseling of deaf people. They describe how these actions can be done in a culturally sensitive way that privileges neither deafness nor hearing, and they review two decades of experience with genetic counseling at Gallaudet University. They also provide data about deaf Americans' attitudes toward bearing deaf children and the abortion of fetuses on the basis of hearing status.

Perhaps the most detailed study of attitudes related to genetic issues, however, is Anna Middleton's work done in England, which she summarizes in her contribution "Deaf and Hearing Adults' Attitudes toward Genetic Testing for Deafness." Middleton's large study covered various groups of hearing and deaf people, including among the latter those who would be identified as culturally deaf and those who saw themselves as audiologically deaf but culturally hearing. She found that deaf people are more likely than hearing people to have negative attitudes toward genetic testing, though not for exactly the same reason as Willis suggests when writing about his own experience as a disabled person. Middleton also discovered, without surprise, that hearing people are much more likely than deaf people to consider aborting a fetus because of its likely deafness. Finally, Middleton raises ethical issues that everyone involved in genetic diagnosis and counseling must confront. She comments on both an Australian couple, who screened embryos before implantation and eliminated those carrying two copies of the connexin 26 mutation to avoid having a deaf child, and an American lesbian couple, who sought a deaf donor for artificial insemination to increase the chances of producing a deaf child, suggesting that thoughtful approaches to genetic engineering will not produce social consensus, at least in with respect to deafness.

Shifra Kisch does not directly approach ethical questions but asks what happens when a traditional community with a high rate of genetic deafness encounters scientific explanations for the anomaly and is offered medically based intervention strategies. She analyzes the language used by a small Bedouin community in the Negev desert where this has happened. She finds that the group she has studied explains deafness in complex and sometimes contradictory ways as they reconcile traditional beliefs with scientific demands. Kisch's discussion brings to mind both Joseph Murray, whose essay earlier in this volume discussed the attitudes of deaf people in the United States and Europe to deaf-deaf marriages, when she writes that "people in the community consider many factors other than the avoidance of one specific genetic risk in seeking a marriage partner," and Nora Groce, whose study *Everyone Here Spoke Sign Language* described another genetically isolated community with a high prevalence of deafness.

Willis's essay, by contrast with the others, is deeply personally, recounting his experiences and those of his family members with disabilities and genetic testing. Like the deaf people of Middleton's survey and the Bedouin community Kisch examined, he is skeptical about genetic testing related to his blindness (though not related to his heart disease), its value, and whether he wishes to participate—to be a guinea pig, in his words—in research studies. Like many culturally deaf people and others with disabilities, Willis writes that he does "not experience vision loss as a disease. It is a different way of perceiving the world." Rather than focus on medical interventions or genetic fixes, he writes, "I think of myself as socially blind; the deficits associated with my blindness result more from society's limitations than from a disease process active in my body." One might conclude, therefore, that social investment in genetic studies should be tempered with investment in accommodations for a diverse citizenry.

GENES FOR DEAFNESS AND THE GENETICS PROGRAM AT GALLAUDET UNIVERSITY

Kathleen S. Arnos and Arti Pandya

At least 1 in 1,000 newborn infants has severe to profound deafness, and genetic factors cause at least 50–60 percent of these cases (Marazita et al. 1993). Hereditary deafness is not a single entity; more than 400 forms are known to exist. Given that there are fewer than 40,000 genes in humans (International Human Genome Sequencing Consortium 2001), the fact that more than 1 percent of all genes are involved in determining the structure and functioning of the ear documents the complexity of this organ. Despite this striking heterogeneity of hereditary deafness, one form, known as *GJB2*, a gene that encodes the protein connexin 26, accounts for 20–30 percent of deafness in the United States (Cohn et al. 1999; Pandya et al. 2003).

The many forms of genetic deafness can sometimes be distinguished from one another by audiologic characteristics, such as the type, degree, or progression of the hearing loss; the vestibular characteristics (the presence or absence of balance problems); the specific mode of inheritance; or the presence or absence of other medical or physical characteristics. Two-thirds of hereditary deafness occurs as an isolated finding, referred to as nonsyndromic deafness. Syndromic deafness, which accounts for the remaining one-third of genetic forms, refers to deafness that is associated with other medical or physical features (Gorlin et al. 1995). For example, some syndromic forms of deafness are associated with serious medical complications, such as alterations of the structure of the kidneys, irregular heart rhythm, and progressive loss of vision. Other syndromic forms of deafness are associated with mild variations, such as pigmentary changes causing two different colored eyes or changes in the hair color, or a thickening of the skin on the palms of the hands and soles of the feet. A few of the more common syndromic forms of deafness will be described later in this chapter.

IDENTIFICATION AND CHARACTERIZATION OF GENES FOR DEAFNESS

More than 110 genes for nonsyndromic and syndromic deafness have been identified in the past decade (Van Camp and Smith 2003). The methodologies used to accomplish this fall under the category of genetic mapping, which is the localization of a gene on a particular region of a chromosome. There are a variety of approaches to genetic mapping (Keats 2000; Avraham 2003), one of which is the study of the transmission of genetic markers through three to four generations of large families with the same form of hereditary deafness. Other techniques take advantage of mouse models of specific forms of deafness or inbred population isolates with several members who are deaf. Once a region is mapped, it is often possible to identify genes in the region and fully characterize them in terms of their biochemical components. It may then be possible to determine the protein product for which the gene codes and how the protein functions in the body (genes code for proteins, which perform a variety of critical functions in the body). This type of gene characterization can also lead to the ability to test for alterations or mutations of the gene. While genetic testing is costly and not now widely available, testing for a few, more common genes will soon become the standard of care for children who are identified as being deaf or hard of hearing (American College of Medical Genetics Expert Panel 2002).

As genes involved in hearing and deafness are discovered, their characterization leads to important insights about the functioning of the intricate components of the inner ear (Tekin et al. 2002). Some of these genes code for proteins that form structural components of the hair cells and membranes of the cochlea. Others code for proteins that are responsible for the transport of ions, such as potassium and sodium, or other molecules, such as calcium, through the cells in the cochlea. The appropriate "balance" of these chemicals in different compartments of the cochlea and their transport from cell to cell are essential to the normal process of hearing. Still other recently discovered genes are important for regulating the development and functioning of the inner ear, including the "turning off" or "turning on" of genes at the appropriate developmental stage. Mutations in any of these genes can result in deafness, both syndromic and nonsyndromic forms.

COMMON SYNDROMIC FORMS OF HEREDITARY DEAFNESS

Usher syndrome refers to a group of several disorders that have in common the features of deafness and retinitis pigmentosa, a progressive degenerative disease of the eye leading to night blindness and tunnel vision (Keats and Corey 1999). All forms of Usher syndrome are inherited as autosomal recessive traits and are classified into three different clinical types distinguished by the presence or absence of changes in the vestibular (balance) system and differences in the progression or severity of the hearing loss. More than eleven different genes are thought to be responsible for the various types of Usher syndrome (Van Camp and Smith 2003). Type I Usher syndrome is characterized by congenital, profound sensorineural deafness, retinitis pigmentosa with onset before puberty, and severe vestibular (balance) problems. At least seven genes are thought to cause Type I Usher syndrome (Van Camp and Smith 2003). The most common of these genes codes for the myosin VIIa protein. This protein forms an important structural component of the hair cells in the ear and also functions in the eye. Mutations in this gene (*MYO7A*) occur in about 60 percent of individuals with Type I Usher syndrome (Kimberling 2003). Type II Usher syndrome is characterized by early-onset, moderate to severe hearing loss; later-onset retinitis pigmentosa; and normal vestibular function. Three genes have been identified for Type II Usher syndrome. Finally, Type III Usher syndrome is a rare form that involves retinitis pigmentosa and progressive hearing loss. Thus far, one gene for this form of Usher syndrome has been identified. The identification and characterization of genes for Usher syndrome offer hope for a better understanding of the devastating visual loss with the future possibility of improved treatments through medications or gene transfer technologies.

Waardenburg syndrome (WS) constitutes a group of genetic disorders that involve pigmentary changes, such as different colored eyes, white streaks of hair, and white patches of skin (Gorlin et al. 1995). There are two more common types of WS (Newton and Read 2003). While the spectrum of pigmentary changes is the same for both forms, Type I WS is characterized by the appearance of wide-spaced eyes, a condition known as dystopia canthorum. Dystopia canthorum does not occur in Type II WS. Both types of WS are inherited as autosomal dominant traits, with variable expression of the features. About 20–25 percent of individuals with Type I WS are deaf in one or both ears, while deafness occurs in about 50 percent of individuals with Type II WS. For both types, the deafness is sensorineural and most often severe to profound in degree. Type III and Type IV WS are rare disorders; Type III WS is associated with musculo-skeletal malformations, whereas Type IV WS is characterized by features of WS II plus a blockage of the intestines known as Hirschprung disease (Newton and Read 2003). Type I WS is caused by mutations in the *PAX3* gene on chromosome 2 (Tassabehji et al. 1992). Mutations in at least two genes are known to cause WS Type II (Van Camp and Smith 2003).

Pendred syndrome is an autosomal recessive condition characterized by sensorineural hearing loss and enlargement of the thyroid gland (goiter), which occurs during the teenage years or early adulthood (Reardon et al. 1997). This syndrome is estimated to occur in up to 10 percent of deaf children. Most individuals with Pendred syndrome have normal thyroid function, even though they may have an enlarged thyroid gland. Structural changes of the inner ear that can be diagnosed with magnetic resonance imaging (MRI) or computerized tomography (CT) scan are common as well. The structural changes can include enlarged vestibular aqueduct (EVA) and/or Mondini dysplasia, a condition in which one of the turns of the cochlea is missing. A gene known as *SLC26A4*, located on chromosome 7, has been identified, and mutations of this gene are seen in many, but not all, individuals who have Pendred syndrome (Scott et al. 1999).

Jervell and Lange-Nielsen (JLN) syndrome is associated with profound, congenital deafness and is also inherited as an autosomal recessive trait. While it is not common, identification of this syndrome early in life is important. The syndrome is associated with the sudden onset of fainting spells due to a defect in the conduction activity of the heart (Gorlin et al. 1995). This heart irregularity can lead to sudden death in children. Early identification by screening for the heart defect using an electrocardiogram (EKG) or a more complete monitoring of the heart over time through a Holter monitor is essential. This heart condition is characterized by an elongation of the QT interval seen on the EKG; the risk of death can be greatly reduced with the use of medication or pacemakers to regulate the heartbeat. Mutations in two genes, *KCNQ1* on chromosome 11 and *KCNE1* on chromosome 21, can cause this syndrome (Neyroud et al. 1997; Schulze-Bahr et al. 1997). The proteins made by these two genes are responsible for the movement of potassium ions into and out of the cells of the heart as well as the ear.

COMMON FORMS OF NONSYNDROMIC DEAFNESS

More than thirty genes for nonsyndromic deafness have been identified. Of these, genes for several connexin proteins have an important role in the development and functioning of the inner ear. The *GJB2* (gap junction beta 2) gene, which codes for the protein connexin 26, was first described in 1997 (Denoyelle et al. 1997; Kelsell et al. 1997). Mutations in this gene are an important cause of congenital, severe-to-profound deafness. It has been estimated that the *GJB2* gene (connexin 26) accounts for 20–30 percent of congenital deafness in the United States (Cohn et al. 1999; Pandya et al. 2003), up to 50 percent of congenital deafness in Spain (Estivill et al. 1998), and a very high, but undetermined, percentage of deafness in the Ashkenazi Jewish population (Morrell et al. 1998). The connexin 26 protein forms a gap junction in the cells underlying the hair cells. To date, more than ninety recessive mutations and a handful of dominant mutations of the *GJB2* gene that cause deafness have been identified (Rabionet et al. 2003). One particular mutation, known as 35delG, is the most common and is found in about 70 percent of all individuals who have *GJB2*-related deafness. Another mutation, 167delT, is common in individuals of Ashkenazi Jewish ethnicity (Morrell et al. 1998).

Most children who become deaf from alterations in connexin 26 have hearing parents, but the gene is also an important cause of deafness in families where deaf parents have all deaf children. When deaf parents both have recessive mutations in connexin 26, all of their children are expected to be deaf. Connexin 26 deafness, as Walter Nance has pointed out, seems to have made an important contribution to the composition of the Deaf community in the United States during the twentieth century.

The *GJB6* gene (connexin 30) is a newly discovered gene (del Castillo et al. 2002) that also causes deafness. A large deletion of a portion of this gene, when it occurs in a double dose (recessive inheritance), has been associated with the occurrence of deafness in a few cases (del Castillo et al. 2002; Pandya et al. 2003). Individuals who inherit one copy of a change in a connexin 26 gene along with one copy of the connexin 30 deletion are also deaf. This is one of the few known examples of a situation where mutations in two different genes cause deafness when inherited together, although the exact mechanism causing the deafness is unknown.

IMPLICATIONS OF GENETIC TESTING

An increasing number of families will have access to genetic testing in the years to come due to more widely available clinical testing for genetic forms of deafness at commercial and private laboratories, as well as increased awareness among health professionals and consumers regarding the benefits of such testing. There are many possible implications of the availability of genetic testing for hereditary deafness, including clinical benefits, ethical considerations, and the availability of adequate and competent genetic counseling (Arnos 2003), as well as the potential impact of such testing on the Deaf community (Jordan 1991; Arnos 2002).

Early diagnosis of the cause of deafness has many benefits for children and their families. Through competent genetic counseling, parents can understand the genetic mechanism causing the deafness and the chance that deafness might recur in future children. This can allow them to make informed choices about their family planning. In some cases, expensive and invasive medical tests can be avoided. For example, for those children in whom connexin 26 is identified as the cause of deafness, other tests commonly done on a deaf child to identify environmental causes of deafness or to rule out syndromes need not be done. Such tests might include titers to identify prenatal infections, tests of heart and thyroid function, and imaging tests of the inner ear. Identification of the exact cause of deafness can also provide information about possible progression of the hearing loss, which may be helpful in making medical and educational choices. When provided by a skilled genetic counselor, knowledge about the cause of deafness may benefit parents, helping to dispel guilt and anxiety and assisting them in moving through the grieving process.

Ethical issues surrounding the introduction and utilization of genetic tests focus on issues such as privacy and confidentiality, informed consent, and discrimination by health insurers or employers based on the results of a genetic test (Secretary's Advisory Committee on Genetic Testing 2003). Genetic testing results can have an impact on the entire family and may place individuals in situations where they find themselves making choices they would have preferred not to make. Risks and benefits of specific types of genetic testing, including testing for genes for deafness, can be explained to families and individuals during the process of genetic counseling.

GENETIC EVALUATION AND COUNSELING FOR DEAF PEOPLE

A genetic evaluation can allow parents of deaf and hard of hearing children and deaf adults themselves to get accurate information about the cause of deafness, other medical implications, and the chance of recurrence in future children. The evaluation process involves the collection of detailed family and medical histories and a thorough physical examination to search for evidence of syndromic forms of deafness (Arnos et al. 1996; Arnos and Pandya 2003). An accurate family history is one of the most important clues to the cause of the deafness. The genetic counselor collects details about the health and hearing status of siblings, parents, grandparents, and other close family members and documents ethnicity, since some forms of genetic deafness are more common in certain ethnic or racial groups. Also, it is determined if the parents may be related to one another. If the parents are consanguineous (blood relatives of one another), this is an important clue regarding the probable inheritance pattern of the deafness.

As part of the evaluation process, the genetics physician determines which types of medical testing are appropriate and then evaluates and interprets any test results. For example, if a specific syndrome is suspected to be present in a deaf child, the physician may recommend tests of thyroid or cardiac function or may refer the child to an ophthalmologist for a complete eye examination. At this point, families who might benefit from genetic testing are informed about the availability of such tests and given appropriate information about the benefits, risks, and implications of such information. Individuals who are to be tested should have a full understanding of all of these aspects and must give informed consent. In most situations, this testing is done by obtaining a small blood sample from which the DNA can be extracted and tested for certain genes.

GENETIC COUNSELING IN THE DEAF COMMUNITY

Genetics professionals increasingly recognize the need to make genetic counseling and evaluation services more culturally sensitive and inclusive of different ethnic and racial groups. Since genetic counselors value nondirective counseling, an understanding of cultural values, beliefs, and practices that could potentially influence the genetic counseling situation is important to ensure respectful behavior and communication between genetics professionals and their clients (Fisher 1994). Religious differences and cultural values can have a profound influence upon what a family will deem acceptable in terms of treatment options or prenatal diagnosis. The role of the family and paternal and maternal roles also vary from culture to culture and can affect choices as well as the most effective way to present genetics information. Some genetic diseases are more common or may occur exclusively in certain ethnic or racial groups, and geneticists have developed strategies to most effectively implement screening and treatment programs for conditions such as Tay Sachs disease, sickle-cell anemia, and thalassemia. The curricula of genetic counseling training programs now place heavy emphasis on training in culturally appropriate genetic counseling.

The importance of making genetic counseling and evaluation services more culturally appropriate for the Deaf community was first addressed by the establishment of the Genetics Program at Gallaudet University in 1984 with funding from the U.S. Department of Health and Human Services (Arnos et al. 1992). A major goal of establishing genetic services at Gallaudet was to develop materials and techniques to make genetic services more accessible to Deaf individuals and to disseminate this model to other genetics centers. The model of genetic counseling that was developed emphasized the use of culturally neutral materials and terms, the development of appropriate visual materials to explain basic genetic concepts that were specific to deafness, and the utilization of appropriate communication for face-to-face contact with Deaf individuals in a genetic evaluation and counseling setting. This included the use of sign language interpreters and development of appropriate signs for genetics terminology. In addition to emphasizing culturally neutral word choice, the importance of nondirective counseling was recognized. For example, genetic counselors should not make assumptions about the desire of Deaf couples to have deaf or hearing children, and they should respect the right of Deaf couples to make such choices (Arnos et al. 1992).

In addition to genetic counseling, the program planners recognized that a strong educational component was necessary to educate the Deaf community about the potential benefits of genetic evaluation and counseling and to educate the medical community, including geneticists and audiologists, about making genetic services accessible to Deaf people. Numerous efforts were made to provide information to health professionals at conferences and meetings, and a number of articles in genetics, audiology, and otolaryngology journals addressed these issues. In 1994, a manual to be used in genetic counseling training programs was developed by a genetic counselor at Gallaudet University, together with many Deaf and hearing professionals who work in the field of deafness (Israel 1995). The authors included information about language and communication; the education of deaf children; and the cultural, psychosocial, demographic, epidemiologic, and genetic aspects of deafness. This manual was distributed to all genetic counseling training programs and continues to be used as part of the curriculum in many of these programs.

GENETICS RESEARCH IN THE DEAF COMMUNITY

Identification of Genes for Deafness

Genetics research in the Deaf community performed collaboratively by the genetics programs at Gallaudet University and the Virginia Commonwealth University has focused on the molecular aspects of various forms of genetics deafness and the attitudes of Deaf and hard

TABLE 1. DISTRIBUTION OF PARTICIPANTS BY FAMILY STRUCTURE

Hearing Status of Family Members	No.	%
Both parents and all siblings hearing	382	51.8
Both parents hearing, one or more siblings deaf	198	26.9
One parent deaf	55	7.5
Both parents deaf	102	13.8
Total	737	100

of hearing people in the United States toward genetics technologies. In 1997, genetic testing for genes for nonsyndromic deafness was offered to members of the Deaf community and the parents of deaf children across the country as part of a research protocol (Pandya et al. 2003). This collaborative group previously conducted molecular genetic studies of WS (Reynolds et al. 1995). The Gallaudet Research Institute (GRI) was key to the successful recruitment of study participants. GRI's Annual Survey of Deaf and Hard of Hearing Children and Youth identified the most appropriate potential research subjects. Participation was then offered to the parents of deaf and hard of hearing children who attend special education programs across the country.

Testing for the connexin 26 gene was part of the research protocol in the recent study on nonsyndromic deafness. Table 1 summarizes data from this testing (Pandya et al. 2003). The data includes individuals from the Deaf community who participated in genetic evaluation and counseling at Gallaudet University, but the majority of individuals tested lived throughout the United States and were contacted through GRI's Annual Survey. Table 1 shows data for 737 deaf people, including one person from each family, although several people in the family, including parents, brothers and sisters, may have been tested. About half the individuals who participated were the only deaf people in their families. Approximately 27 percent had hearing parents but a deaf sibling. Seven percent had one deaf parent, and 14 percent had two deaf parents.

Table 2, summarized from Pandya et al. (2003), shows the results of testing for the connexin 26 gene in these same individuals. Overall, 22 percent were deaf from the connexin 26 gene. Some interesting

TABLE 2. FREQUENCY OF CONNEXIN 26 DEAFNESS BY
           FAMILY STRUCTURE

Hearing Status of Family Members	*n* = 737
	No. (%)
Both parents and all siblings hearing	52 (13.6)
Both parents hearing, one or more siblings deaf	64 (32.3)
One parent deaf	4 (7.3)
Both parents deaf	43 (42.2)
All tested individuals	163 (22.1)

comparisons can be made when this is broken down by type of family structure. For example, when there was only one deaf person in the family, the connexin 26 gene accounted for about 14 percent of the instances of deafness. In deaf persons with one or more deaf siblings and hearing parents, however, the same gene caused 32 percent of the instances of deafness. Finally, in those families where both parents were also deaf, 42 percent of those tested were found to be deaf from connexin 26. This data documents the importance of the connexin 26 gene as a cause for deafness within the American Deaf community.

Attitudes of Deaf and Hard of Hearing People toward Genetics

Based on the model provided by Anna Middleton's studies of attitudes toward genetics among deaf and hard of hearing people in the United Kingdom, the genetics programs at the Medical College of Virginia and Gallaudet University collaborated to study the attitudes of the Deaf community in the United States (Stern et al. 2002). In this study, deaf and hard of hearing people affiliated with Self Help for Hard of Hearing People, Inc. (a national organization), the National Association of the Deaf, and the student body of Gallaudet University were sent a questionnaire assessing attitudes about genetic testing and prenatal diagnosis. This survey documented that, overall, deaf and hard of hearing people in the United States have a positive attitude toward genetics, have no preference about the hearing status of their children, are not interested in prenatal diagnosis for hearing status, and think pregnancy termination for hearing status should be illegal. These data were further analyzed by dividing responses according to whether the respondents identified themselves more with the Deaf community, more with the hearing community, or with both communities equally, which revealed some striking differences among the groups.

Although 26.6 percent of culturally Deaf individuals preferred deaf children, the majority of this group and of those with equal involvement in both cultures had no preference for the hearing status of their children. Statistically significant differences existed between the groups on the issue of abortion. The majority of respondents in each group felt that abortion of a deaf baby when a hearing baby is preferred should be illegal. However, 8 percent of the culturally hearing group and 2.1 percent of those with equal involvement in both cultures would consider aborting a deaf baby. No culturally Deaf respondents said they would abort a deaf baby. Similarly, the majority of respondents in all three groups felt that abortion of a hearing baby when a deaf baby is preferred should be illegal. However, 2.7 percent of the culturally Deaf group, 0.8 percent of the culturally hearing group, and 2.1 percent of the equal involvement group would consider abortion of a hearing baby. Further studies of this issue as well as the potential use of genetic testing results for the purpose of mate selection to either avoid or ensure the birth of deaf children are planned.

SUMMARY AND CONCLUSIONS

Genetic factors are a significant cause of deafness, accounting for more than half of the instances of children who are either born deaf or become deaf early in life. Recent developments in molecular genetics have allowed the identification of dozens of genes for deafness. Genetic testing is now considered to be an important component of the diagnostic process for individuals who are identified as being deaf. Genetic evaluation and counseling, in addition to genetic testing, can have many benefits for deaf and hard of hearing individuals and their families. The accurate identification of the cause of deafness, either genetic or nongenetic, can have important implications for educational assessment and the psychological well-being of the family, including their acceptance of and adjustment to the deafness. For deaf and hard of hearing adults, information regarding the cause of deafness can be empowering, enabling them to understand their own health status and make informed choices regarding the hearing status of their children.

The genetics program at Gallaudet University, which has served the Deaf community for almost twenty years, provides a model for culturally sensitive genetic counseling for Deaf people. Geneticists at Gallaudet, in collaboration with colleagues at Virginia Commonwealth University, have also studied the ethical implications of genetic testing for deafness by examining the attitudes of deaf and hard of hearing people and the parents of deaf children in the United States toward the use of genetics technology. These studies have shown that while these individuals are generally interested in the use of genetic testing to learn more about their or their child's cause of deafness, there is little support for the use of prenatal genetic testing for the purpose of selection of the hearing status of the child. These and future studies may continue to shed light on the potential use of genetic technology by individuals and the possible impact on the Deaf and hearing communities.

REFERENCES

American College of Medical Genetics Expert Panel. 2002. Genetics evaluation guidelines for the etiologic diagnosis of congenital hearing loss. *Genetics in Medicine* 4(3): 162–71.

Arnos, K. S. 2002. Genetics and deafness: Impacts on the Deaf community. *Sign Language Studies* 2(2): 150–68.

———. 2003. The implications of genetic testing for deafness. *Ear & Hearing* 24(4): 324–31.

Arnos, K. S., M. Cunningham, J. Israel, and M. Marazita. 1992. Innovative approach to genetic counseling services for the deaf population. *American Journal of Medical Genetics* 44:345–51.

Arnos, K. S., J. Israel, L. Devlin, and M. P. Wilson. 1996. Genetics aspects of hearing loss in children. In *Hearing Care in Children*, ed. J. Clark and F. Martin, 20–44. Needham Heights, Mass.: Allyn & Bacon.

Arnos, K. S., and A. Pandya. 2003. Advances in the genetics of deafness. In *Handbook of Deaf Studies, Language and Education*, ed. M. Marschark and P. Spencer, 392–405. New York: Oxford University Press.

Avraham, K. B. 2003. Mouse models for deafness: Lessons for the human inner ear and hearing loss. *Ear & Hearing* 24(4): 332–41.

Cohn, E. S., P. M. Kelley, T. W. Fowler, M. P. Gorga, D. M. Lefkowitz, H. J. Kuehn, G. B. Schaefer, L. S. Gobar, F. J. Hanh, D. J. Harris, and

W. J. Kimberling. 1999. Clinical studies of families with hearing loss attributable to mutations in the connexin 26 gene (*GJB2/DFNB1*). *Pediatrics* 103:546–50.

del Castillo, I., M. Villamar, M. A. Moreno-Pelayo, F. J. del Castillo, A. Alvarez, D. Telleria, I. Menendez, and F. Moreno. 2002. A deletion involving the connexin 30 gene in nonsyndromic hearing impairment. *New England Journal of Medicine* 346(4): 243–49.

Denoyelle, F., D. Weil, M. A. Maw, S. A. Wilcox, N. J. Lench, D. R. Allen-Powell, A. H. Osborn, H-HM. Dahl, A. Middleton, M. J. Houseman, C. Dode, S. Marlin, A. Boulila-ElGaied, M. Grati, H. Ayadi, S. BenArab, P. Bitoun, G. Lina-Granade, J. Godet, M. Mustapha, J. Loiselet, E. El-Zir, A. Aubois, A. Joannard, R. J. McKinlay Gardner, and C. Petit. 1997. Prelingual deafness: High prevalence of a 30delG mutation in the connexin 26 gene. *Human Molecular Genetics* 6:2173–177.

Estivill, X., P. Fortina, S. Surrey, R. Rabionet, S. Melchionda, L. D'Agruma, E. Mansfield, E. Rappaport, N. Govea, M. Mila, L. Zelante, and P. Gasparini. 1998. Connexin-26 mutations in sporadic and inherited sensorineural deafness. *Lancet* 351:394–98.

Fisher, N. L. 1996. *Cultural and Ethnic Diversity: A Guide for Genetics Professionals.* Baltimore: Johns Hopkins University Press.

Gorlin, R. J., H. V. Torielo, and M. M. Cohen. 1995. *Hereditary Hearing Loss and Its Syndromes.* New York: Oxford University Press.

International Human Genome Sequencing Consortium. 2001. Initial sequencing and analysis of the human genome. *Nature* 409:860–921.

Israel, J. N. 1995. *An Introduction to Deafness: A Manual for Genetic Counselors.* Washington, D.C.: Gallaudet University Research Institute.

Jordan, I. K. 1991. Ethical issues in the genetic study of deafness. *Annals of the New York Academy of Sciences* 630:236–39.

Keats, B. J. B. 2000. Genetic intervention and hearing loss. In *Audiology*, ed. R. J. Roeser, M. Valente, and H. Hosford-Dunn, 593–614. New York: Thieme.

Keats, B. J. and D. P. Corey. 1999. The Usher syndromes. *American Journal of Medical Genetics* 89:158–66.

Kelsell, D. P., J. Dunlop, H. P. Stevens, N. J. Lench, J. N. Liang, G. Parry, R. F. Mueller, and I. M. Leight. 1997. Connexin 26 mutations in hereditary non-syndromic sensorineural deafness. *Nature* 387:80–83.

Kimberling, W. J. 2003. Clinical and genetic heterogeneity of Usher syndrome. *Audiological Medicine* 1:67–70.

Marazita, M. L., L. M. Ploughman, B. Rawlings, E. Remington, K. S. Arnos. and W. E. Nance. 1993. Genetic epidemiological studies of early-onset deafness in the U.S. school-age population. *American Journal of Medical Genetics* 46:486–91.

Morrell, R. J., H. J. Kim, L. J. Hood, L. Goforth, K. Friderici, R. Risher, G. Van Camp, C. I. Berlin, C. Oddoux, H. Ostrer, B. Keats, and T. B. Friedman. 1998. Mutations in the connexin 26 gene (*GJB2*) among Ashkenazi Jews with nonsyndromic recessive deafness. *New England Journal of Medicine* 339:1500–505.

Newton, V. E., and A. P. Read. 2003. Waardenburg syndrome. *Audiological Medicine* 1:77–88.

Neyroud, N., F. Tesson, I. Denjoy, M. Leibovici, C. Donger, J. Barhanin, S. Faure, F. Gary, P. Coumel, C. Petit, K. Schwartz, and P. Guicheney. 1997. A novel mutation in the potassium channel gene *KVLQT1* causes the Jervell and Lange-Nielsen cardioauditory syndrome. *Nature Genetics* 15:186–89.

Pandya, A., K. S. Arnos, X. J. Xia, K. O. Welch, S. H. Blanton, T. B. Friedman, G. G. Sanchez, X. Z. Liu, R. Morrell, and W. E. Nance. 2003. Frequency and distribution of *GJB2* (connexin 26) and *GJB6* (connexin 30) mutations in a large North American repository of deaf probands. *Genetics in Medicine* 5(4): 295–303.

Rabionet, R., P. Gasparini, and X. Estivill. Connexin homepage. http://www.crg.es/deafness (accessed April 2003).

Reardon, W., R. Coffey, P. D. Phelps, L. M. Luxon, D. Stephens, P. Kendall-Taylor, K. E. Britton, A. Grossman, and R. Trembath. 1997. Pendred syndrome—100 years of underascertainment? *Quarterly Journal of Medicine* 90:443–47.

Reynolds, J. E., K. S. Arnos, B. Landa, C. A. Stevens, B. A. Salbert, L. Wright, B. Duke, W. Hunt, M. L. Marazita, L. Ploughman, C. MacLean, W. E. Nance, and S. R. Diehl. 1995. Analysis of locus heterogeneity in Waardenburg syndrome types 1 and 2 using highly informative microsatellite markers. *Human Heredity* 45:243–52.

Schulze-Bahr, E., Q. Wang, H. Wedekind, W. Haverkamp, Q. Chen, Y. Sun, C. Rubie, M. Hordt, J. A. Towbin, M. Borggrefe, G. Assmann, X. Qu, J. C. Somberg, G. Breithardt, C. Oberti, and H. Funke. 1997. KCNE1 mutations cause Jervell and Lange-Nielsen syndrome. *Nature Genetics* 17:267–68.

Scott, D. A., R. Wang, T. M. Kreman, V. C. Sheffield, and L. P. Karnishki. 1999. The Pendred syndrome gene encodes a chloride-iodide transport protein. *Nature Genetics* 21:440–43.

Secretary's Advisory Committee on Genetic Testing. http://www4.od.nih.gov/oba/sacgt/GTDocuments.html (accessed April 2003).

Stern, S. J., K. S. Arnos, L. Murrelle, K. O. Welch, W. E. Nance, and A. Pandya. 2002. Attitudes of deaf and hard of hearing subjects towards genetic testing and prenatal diagnosis of hearing loss. *Journal of Medical Genetics* 39:449–53.

Tassabehji, M., A. P. Read, V. E. Newton, R. Harris, R. Balling, P. Gruss, and T. Strachan. 1992. Waardenburg's syndrome patients have mutations in the human homologue of the Pax-3 paired box gene. *Nature* 355:635–36.

Tekin, M., K. S. Arnos, and A. Pandya. 2001. Advances in hereditary deafness. *Lancet* 358:1082–90.

Van Camp, G., and R. J. H. Smith. Hereditary Hearing Loss Homepage. http://dnalab-www.uia.ac.be/dnalab/hhh/ (accessed April, 2003).

DEAF AND HEARING ADULTS' ATTITUDES TOWARD GENETIC TESTING FOR DEAFNESS

Anna Middleton

Genetic factors play a major role in the development of both congenital and late-onset deafness (Cohen and Gorlin 1995). More than 120 different genetic loci involved with deafness have been identified over the past ten years (Van Camp and Smith 2004), and one particular gene, *GJB2* or connexin 26, is thought to play a part in the most common form of genetic deafness—nonsyndromal recessive deafness. This is deafness in isolation (not part of a syndrome), and the person who is deaf usually has two parents who are both hearing but carriers of an altered gene, such as connexin 26.

Testing for *alterations* in such a gene can be done via a blood sample. A *diagnostic* genetic test can inform a deaf person if his or her deafness is likely to be due to known genetic factors. A *carrier* genetic test can inform a hearing person if he or she has a deafness-causing gene. If both partners are hearing but are carriers for the same altered gene, they have a one in four chance of having deaf children. A *prenatal* genetic test is a test in pregnancy that can inform a pregnant couple whether their fetus has the gene alterations that could cause it to be deaf (but would not indicate to what level). There are many genetics laboratories that now offer genetic testing for changes in various deafness genes, particularly connexin 26. This chapter considers research that ascertains how deaf people and their families feel about this testing.

DEAFNESS—MEDICAL PROBLEM OR CULTURAL DIFFERENCE?

Deafness can be viewed from different perspectives. People who are culturally Deaf (written with an uppercase "D") may not predominantly perceive their deafness as a problem that needs to be "treated" with a hearing aid or cochlear implant. It is the medical model that would consider deafness in this way. The cultural or sociological model views deafness as a condition to be preserved and celebrated, offering a strong identity, rich language, and a distinct cultural community (Padden 1980). Many Deaf people who embrace this perspective do not want to be treated for their deafness and reject medical services that may offer this. For them personally, their deafness is not a disability; it is societal attitudes that are disabling.

PRENATAL GENETIC TESTING FOR DEAFNESS

Many clinical and research professionals involved with deaf families believe that the incorporation of genetic testing for deafness should be part of routine practice within clinical genetics services (Reardon 1998). If carrier and diagnostic genetic testing become more widely available, then it is almost implicit that prenatal genetic testing could also be on offer. For example, if a deaf child has a diagnostic genetic test that confirms his or her deafness is due to altered connexin 26 genes, and the parents are confirmed as carriers for this genetic alteration, then they know that they have a one in four chance of having more deaf children. Parents may wish to have a prenatal genetic test during a subsequent pregnancy, and if the fetus is found to have two altered connexin 26 genes (and thus is likely to be deaf), they may choose to end the pregnancy. For parents who have had a particularly difficult time with their deaf child(ren) (e.g., in obtaining education or support in teaching their child to communicate), this may well be an option they choose for future pregnancies (Middleton 2004). For culturally Deaf people and also many parents of deaf children, and also deaf, hard of hearing, and deafened adults, this eventuality could be seen very negatively. For them, deafness may not be seen as a condition "serious" enough to warrant an abortion.

An additional dynamic is that some Deaf parents prefer to have deaf children and do not want the numbers of deaf children born to be reduced, threatening the future of their culture (Middleton et al. 1998). Hearing people with no knowledge of Deaf culture may find this perspective difficult to understand.

The following paper details a large research project that documented attitudes and beliefs about genetics and prenatal genetic testing for deafness. The hypothesis was that deaf and hearing people would have different attitudes toward such testing.

The terminology includes using "deaf" to refer to all individuals affected by hearing loss, including the culturally Deaf, and the term "Deaf" to refer to culturally Deaf individuals only.

METHODS

Participants

The study involved 644 deaf individuals, 143 hard of hearing and deafened individuals, and 527 hearing individuals with either a deaf parent or deaf child. The participants determined their own hearing status classifications. Those who termed themselves "deaf" tended to have severe or profound hearing loss; the "hard of hearing" tended to have a mild or moderate hearing loss. "Deafened" usually meant they had lost their hearing later on in life. Sociodemographic data relating to these participants is given in table 1. Culturally Deaf participants were those who considered themselves deaf, hard of hearing, or deafened; said they used British Sign Language (BSL) as their preferred language; and associated more with the Deaf community rather than Hearing World. Therefore, only those participants who met all these criteria collectively were classified as culturally Deaf. Numbers are given in table 2.

Ascertainment

Participants were collected from a number of different sources in the United Kingdom between June 1998 and June 1999. These included various hospital departments such as Clinical Genetics, ENT, and Audiology as well as social services for deaf people, schools, colleges, and

TABLE 1. SOCIODEMOGRAPHIC INFORMATION FOR THE SAMPLE

Sociodemographic Characteristic	Total Sample Size = 1314	%
Age range		
13–19	36	3
20–29	152	12
30–39	441	35
40–49	390	30
50–59	179	14
60–69	54	4
70–93	32	2
Female	833	63
Had Children (deaf or hearing)	958	73
Married or Living with Partner	930	71
Owns own Home	922	70
Had a Religious Affinity	662	50

charities. A letter was sent to each health or education professional involved with the different potential participants, asking them to pass on a questionnaire to their clients. Questionnaires were also handed out at deaf clubs and to delegates attending conferences for deaf people. Participants were given the option of completing the questionnaire via a sign language interpreter if they preferred not to use a written

TABLE 2. NUMBER OF CULTURALLY AND NONCULTURALLY DEAF PARTICIPANTS

All participants with a hearing loss	664	%
—Culturally Deaf	212	32
—Nonculturally deaf	452	68

format. The questionnaire was also posted directly to participants as part of their subscription to three different magazines for deaf people. Recipients were asked to return the completed questionnaire if they wished to take part in the study. St. James's University Hospital Ethics Committee (Leeds, U.K.) granted ethical approval for the research.

Questionnaire

Deaf sign language users provided input to ensure that the questionnaire design was Deaf-sensitive and easily translated into BSL. For example, in the questionnaire, the term *abortion* rather than *termination of pregnancy* was used, since the former translated more fluently into BSL. For Deaf readers, this term is also used in the written text here.

This chapter discusses a small selection of the questionnaire's twenty-one questions, which covered such issues as preference for having deaf or hearing children, interest in utilizing a prenatal genetic test for deafness, reasons for having such testing for deafness, and interest in abortion for deafness and "hearingness" (i.e., ending the pregnancy if the fetus is found NOT to have the deafness-causing genes, since the parents prefer to have deaf children).

RESULTS

Questionnaires were made available to 6,674 potential participants, and 1,314 were returned (average response rate: 20%). The following sections give details of responses to specific questions.

Feelings about New Discoveries in Genetics

Participants were given a list of positive, neutral, and negative words and asked to check off those that described their feelings about new discoveries in genetics. The results showed very different attitudes between groups (figure 1). Deaf participants were more likely to select negative words ($\chi^2 = 42.2$, $df = 6$, $P < 0.001$). The most popular word was *concerned*. Hearing participants were more likely to select positive words ($\chi^2 = 156.7$, $df = 8$, $P < 0.001$). The most popular word was *hopeful*. Hard of hearing and deafened participants were more likely to check a mixture of words. The most popular was *cautious*. This latter group's attitudes reflected the whole spectrum of views—some had the same as

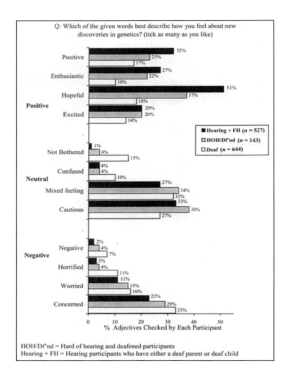

Figure 1. Percentage of participants checking different adjectives to describe their feelings about new discoveries in genetics

the deaf group, some the same as the hearing group, and many in between.

Participants were given the opportunity to comment on their feelings about new discoveries in genetics. The following are a selection of these.

Some participants felt that new discoveries in genetics would be positive:

> We must go forward in genetics to help us understand causes of deafness and other disabilities caused through genes.
>
> *A nonculturally deaf participant*

> I think it is a good idea—to stop the genes passing on into the next generation.
>
> *A nonculturally deaf participant*

Some had negative comments about new discoveries in genetics:

> Angry at people trying to mess with nature and interfering with deaf people—leave us alone!
>
> *A culturally Deaf participant*

> "My hands is little nerve [I feel nervous]. To think it is worst soon [I feel this is the worst situation].
>
> *A culturally Deaf participant, who used BSL as their first language, translated his or her feelings from BSL into written English*

And some comments were mixed:

> Interested but do not feel involved.
>
> *A nonculturally deaf participant*

> Enthusiastic about benefits it can bring—early diagnosis, treatment to improved levels/quality of hearing, BUT concerned it will be used to increase abortion.
>
> *A hearing parent of deaf children*

Interest in Using Prenatal Genetic Diagnosis for Deafness

Deaf participants were less interested in prenatal diagnosis for deafness than hearing participants (χ^2 = 113.1, df = 4, P < 0.0001; see figure 2). Out of the deaf participants, those who were culturally Deaf were the least likely to be interested in prenatal diagnosis for deafness (χ^2 = 21.1, df = 2, P < 0.0001; see figure 3).

Those participants who said that they would be interested in prenatal diagnosis for deafness (figure 2) were also asked if they would prefer to have deaf or hearing children (or did not mind). Therefore the data from figure 2 is presented again to show which participants would have a prenatal genetic test and whether they would prefer to have deaf or hearing children. The results of this are presented in figure 4, and they show that while 69% of hearing and 61% of hard of hearing and deafened participants might use prenatal genetic diagnosis for deafness to have hearing children, deaf participants were not so emphatic. Of those who wanted prenatal genetic diagnosis for deafness, 56% did not mind

Anna Middleton

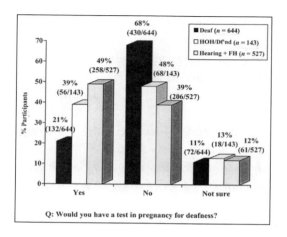

Figure 2. Percentage of participants who were interested in a prenatal genetic test for deafness

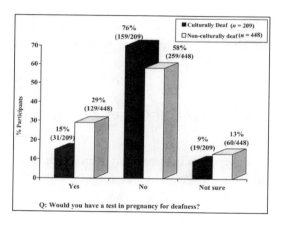

Figure 3. Percentage of culturally Deaf and nonculturally deaf participants who were interested in a prenatal genetic test for deafness

whether their future children were deaf or hearing. For the deaf participants who did mind, 36% said they would rather have hearing children, and 8% said they would rather have deaf children ($\chi^2 = 41.3$, $df = 2$, $P < 0.0001$). Out of the deaf participants who would have a prenatal genetic test and who preferred to have deaf children, the majority (nine out of eleven, i.e., 82%) were culturally Deaf.

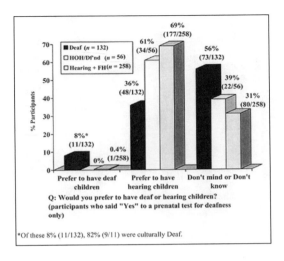

Figure 4. Percentage of participants who wanted prenatal genetic testing and preferred to have a child of a specific hearing status

Reasons for Having Prenatal Genetic Diagnosis for Deafness

Participants were asked to give their reasons for wanting to have prenatal genetic diagnosis. Figure 5 shows that the majority of all groups would only use such testing for preparation—so they could prepare mentally or so that they could prepare for the language needs of the child. The majority would not choose to have an abortion if the fetus was found to have a hearing status the parents did not want. However, a small number of participants (6% deaf, 11% hard of hearing and deafened, 16% hearing) said they would be prepared to consider an abortion if the fetus was deaf; 2% of the deaf participants also said they would consider having an abortion if the fetus was found to be hearing (since they preferred to have deaf children). Of these 2% (three individuals), two were culturally Deaf, and one did not identify with the Deaf community.

Participants were asked to give their comments on abortion for the "wrong" hearing status (be that deaf or hearing depending on what the parents prefer).

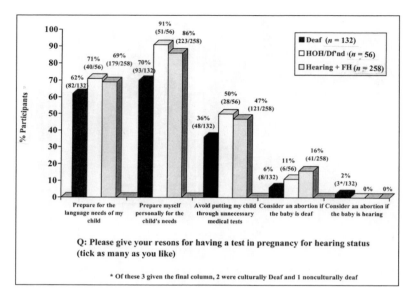

Figure 5. Reasons participants gave for having prenatal genetic testing for hearing status

One participant who would have an abortion for deafness commented:

> I would not wish deafness on my worst enemy. I certainly would not on my own child if it could be avoided.
>
> *A nonculturally deaf participant*

Another participant who would have an abortion for a hearing fetus stated:

> [I would prefer to have deaf children since] I worry that a hearing child would not learn speech and be taken away from me by social services.
>
> *A nonculturally deaf participant*

Participants who would not have an abortion for the "wrong" hearing status made several comments:

I will be disappointed if I'll have deaf babies, but I'll have to accept it. (Deaf culture will hate me if they read that!)

A culturally Deaf participant

I think it's disgraceful to even consider an abortion of a deaf baby. My hearing friends are horrified.

A nonculturally deaf participant

Attitudes to deafness change once a baby is born. You might agree with having an abortion, but once you have a deaf child, you realize what a positive experience it really is.

A hearing participant

Abortion is a personal issue which differs for every individual, However, I strongly believe that more time and effort should be spent on attempting to integrate deaf people into society instead of being treated as a minority.

A nonculturally deaf participant

DISCUSSION

This study documented the attitudes of deaf, hard of hearing, deafened, and hearing participants towards issues surrounding genetics and prenatal genetic testing for deafness.

Hearing participants who had knowledge and experience of deafness through their family history in either their parents or children were more likely (49%) than other groups to feel "hopeful" about new discoveries in genetics and also be interested in having a prenatal genetic test for deafness (compared with 39% hard of hearing and deafened and 21% deaf participants who said they were interested in prenatal genetic testing for deafness). Those hearing, hard of hearing, and deafened participants who were interested in prenatal genetic testing were more likely to prefer to have hearing children. Therefore one could assume that they may choose to end the pregnancy of a deaf fetus. Alternatively, they may accept the baby if it is deaf but might initially be extremely disappointed. They may use the information gained from a prenatal genetic test to prepare themselves for their baby and allow time to come to terms with their situation. Such participants

may have this attitude because they view deafness more from a negative perspective.

Deaf children tend to have more behavioral problems than other
children (Meadow 1976), and hearing parents of deaf children can
often find it difficult to raise a deaf child (Schum 1991). It is understandable, therefore, that some people with personal experience of deafness may perceive it as a difficult condition with which to cope.

The results from the present study can be compared with findings
from other studies that have documented the views of hearing people
toward this issue. Brunger et al. (2000) studied the attitudes of 96 hearing parents of deaf children ascertained via a hospital setting. They
found that 96% had a positive attitude towards genetic testing, and
87% of these were interested in having prenatal genetic testing for deafness for preparation reasons (none indicated they would have an abortion for deafness).

Another study by Martinez et al. (2003) documented the views of
133 hearing students from a U.S. university that had significant numbers of deaf and hard of hearing students (i.e., these hearing students
were therefore familiar with deafness). They found that 64% of hearing students said that they would be interested in having prenatal genetic
diagnosis for deafness, with no comment on whether they would have
an abortion for deafness. These two papers report a much higher interest in prenatal genetic testing for deafness among hearing people than
the present study. Brunger et al. (2000) also reported that there was
great naiveté and misunderstanding about genetics and inheritance patterns. It is possible therefore that, although an immediate response was
that having a prenatal genetic test for deafness was a preferable option,
accurate counseling and information about what the test could offer could
reduce the actual numbers wanting the test. Since the study participants for Martinez et al. (2003) were students with no family history
of deafness, it could also be argued that the issue of having deaf children and having to contemplate a prenatal genetic test for deafness is
not even relevant, and thus responses to this issue may not reflect a
true situation. A future study looking at pregnant couples where prenatal genetic testing for deafness is a realistic proposition would offer
a useful comparison of results.

Out of all the participants in the present study, deaf participants were
the least interested (21%) in having a prenatal genetic test for deaf-

ness, and of those deaf participants who were interested, the majority did not identify with Deaf culture. This result is very similar to that found by Stern et al. (2002), who used the same study questionnaire. Here the researchers documented the views of deaf (n = 135), hard of hearing, deafened (n = 166), and hearing individuals (n = 37) ascertained from support groups for deaf people, a genetics clinic, and Gallaudet University. This research showed that approximately 23% of deaf participants were interested in prenatal genetic diagnoses for deafness. Such a low interest in prenatal genetic diagnosis for deafness is likely due to deaf people generally not perceiving a problem with having deaf children themselves, and therefore not being interested in knowing prenatally if their child is likely to be deaf. Deaf parents of deaf children are less likely to have the same problems raising their children as hearing parents (Schum 1991). This may be due to shared language and understanding of what being *deaf* means.

Culturally Deaf participants were the least interested in prenatal genetic testing for deafness and were also the most likely to say they felt "concerned" about new discoveries in genetics. These findings fit in with previous research and show that there may also be less interest in a technology that could potentially reduce the numbers of deaf people born because of the direct effect on the viability of the Deaf community (Middleton et al. 1998).

Current prenatal genetic tests are done via chorionic villus sampling or amniocentesis at approximately eleven to sixteen weeks into the pregnancy, where a needle takes cells from the placenta or amniotic fluid, respectively, to be used for the genetic testing. Both are invasive procedures and confer a risk of miscarriage from the procedure itself. Therefore, it would be useful to know if there is the same interest in prenatal genetic testing when this information is given. Would prospective parents still be interested in testing and accept the risk of miscarriage just to know whether their baby is deaf or hearing before it is born?

Deaf Parents Preferring To Have Deaf Children

Out of the deaf participants who said they would have prenatal genetic testing for deafness, 8% said that they would prefer to have deaf children. This may mean that they would consider using prenatal genetic testing with selective abortion of a hearing fetus. When specifically asked about this, 2% said that they would consider this course of action. It is not known, however, if faced with a live pregnancy, whether many deaf people would proceed with this. It is also questionable whether geneticists would support the testing and whether obstetricians would perform the abortion. However, three individuals (2%) felt strongly enough on this subject to suggest they would go ahead with such an action.

The issue of deaf parents preferring to have deaf children is not a new phenomenon; it has been well documented in the past. Passing on deafness to the next generation would keep the Deaf culture alive and would mean that the Deaf community would continue to thrive (Jordan 1991; Dolnick 1993; Middleton et al. 1998). Dolnick (1993) comments on this in "Deafness as Culture."

> So strong is the feeling of cultural solidarity that many deaf parents cheer on discovering that their baby is deaf.

Those deaf participants without ties to Deaf culture who nevertheless preferred to have deaf children may have this view because the thought of having hearing children fills them with fear—leading to difficult questions: "How will I cope?" "How will I teach the child to speak?" "What school will they go to?" Schein (1989) suggests that the psychological reaction of a deaf parent to having a child of unexpected hearing status (either deaf or hearing) is very similar to a hearing parent having a deaf child with potential feelings of disbelief, fear, and grief. If deaf parents already had other deaf children, then it is possible that another deaf child would fit into the family unit more successfully, as opposed to a hearing child who may feel isolated. One hearing participant in the present study indicated that she was from a deaf family with several generations of deafness and that she would actually prefer to have deaf children even though she was personally hearing (Middleton 1999). As a hearing person among many generations of deafness, this participant found it hard to cope with being different from the rest of the family.

Abortion for Hearing Status

The research has shown that although there may be interest in having prenatal genetic testing for deafness, the majority of participants would not want to have an abortion for hearing status, be that deaf or hearing. This finding is also supported in other studies done on this subject (Stern et al. 2002; Brunger et al. 2000).

A minority of participants were interested in abortion for hearing status—16% of the hearing participants said they would have an abortion for deafness, and 2% of the deaf participants (three individuals) said they would consider an abortion for a hearing fetus. The figures are very small, but nevertheless still present. Out of the three deaf participants who said they could have an abortion for a hearing fetus, two of them were culturally Deaf. This finding fits in with the Deaf culture stereotype—that deafness is such a positive experience that people may go to extensive lengths to ensure it is passed on. However, when faced with a live, wanted pregnancy, it is debatable whether deaf couples would choose this option. The decision people think they might make may be very different from the decision they actually make when faced with a real situation.

Therefore, it seems that hearing status is not a "condition" that most people feel is "serious" enough to warrant ending a pregnancy. This will be reassuring news for deaf and hearing people who are particularly concerned that the introduction of more genetic testing for deafness will lead to a significant increase in abortion for deafness.

Preimplantation Genetic Diagnosis for Deafness

One alternative to coping with abortion for hearing status could be the use of preimplantation genetic diagnosis when in vitro fertilization is used to create several embryos. These are then tested to see if they have the altered genes for deafness or not, and those that the parents want are selected and transferred back to the mother. There is no need for abortion and the trauma that this may bring since, in theory, the unaffected embryos have already been implanted. However, the current process of IVF itself can be lengthy and the success rate low, so the emotional burden of this should not be underestimated.

Preimplantation genetic diagnosis has already been accepted for hearing parents who would prefer to select deafness out of their future pregnancies. An Australian couple won the right to use preimplantation genetic diagnosis to ensure that their next baby was not deaf; they were both carriers for alterations in the connexin 26 gene (Kelly 2002;

Australasian Bioethics Information 2002). The couple went ahead with this procedure, but to date, it is thought that this was unsuccessful (Noble 2003).

Bioethicists have declared that offering preimplantation genetic diagnosis to enable hearing parents to avoid having deaf children is discriminatory and that deafness is not a valid reason for such selection (Kelly 2002). The Infertility Treatment Authority (ITA) in Melbourne who sanctioned the use of this technology have said it can be available to enable hearing couples to have hearing children, but the same technology would not be available to a deaf couple wanting to have deaf children (Infertility Treatment Authority 2003). This appears to discriminate against potential Deaf couples who may prefer to have deaf children.

Choosing to Have Deaf Children

Deaf couples may want to use genetic technologies to enable them to have deaf children. In 2002, a deaf lesbian couple from the United States chose to have artificial insemination from a male deaf friend with the hope that this would increase their chances of having a deaf child (McLellan 2002). Although not actively using genetic intervention, they hoped that genetic inheritance would be favorable for them, as they wanted to increase the chances of passing deafness on. This case caused international debate about the ethics of deliberately creating what some people felt was a "disabled" child (Levy 2002; Spriggs 2002; McLellan 2002; Fletcher 2002; Anstey 2002; Savulescu 2002). The following are some comments from these articles:

> Couples who select disabled rather than non-disabled offspring should be allowed to make those choices, even though they may be having a child with worse life prospects. (Savulescu 2002)
>
> Couples should not be allowed to select neither for or against deafness. (Anstey 2002)
>
> Deaf people are behaving like hearing people. They feel good about themselves and want to have babies like them. Why should they be morally blamed? (Fletcher 2002)
>
> To intentionally give a child a disability . . . is incredibly selfish. (Ken Connor, president of the Family Research Council, in Spriggs 2002)

Cultures are simply the kind of things to which we are born, and therefore to which the children of deaf parents, hearing or deaf, normally belong. Thus these parents are making a mistake in choosing deafness for their children. Given their own experience of isolation as children, however, it is a mistake which is understandable, and our reaction to them ought to be compassion, not condemnation. (Levy 2002)

There is currently no worldwide consensus of opinion on whether this prenatal testing for deafness with selective abortion for the "wrong" hearing status should be routinely available or not.

Opinions of Genetics Professionals

The American Medical Association (1994) recognises that genetic technology poses dilemmas, in that it is unclear to what extent parents should be able to externally control the genetic makeup of their children. Best practice within today's genetic counseling services is to offer nondirective information about genetics and help clients make fully informed decisions that are right for them. However, if geneticists and genetic counselors are to be truly nondirective, it should be possible for parents to receive prenatal or preimplantation genetic diagnosis for deafness, whether the parents are deaf or hearing, and whether they wish to select either deaf or hearing children. There is currently no policy that determines how far this nondirective approach would be taken with regard to this issue. There seems to be a general view that this would be self-regulating, meaning that deaf or hearing parents would, on the whole, not be interested in using this technology; those that are would be considered on a case-by-case basis.

It is debatable whether geneticists would feel able to offer such testing. In a large international study of genetics professionals, attitudes towards offering prenatal genetic diagnosis to a deaf couple wanting to have deaf children varied across the world (Wertz and Fletcher 1999). The percentage of genetics professionals who said they would perform prenatal genetic diagnosis for a deaf couple wanting to have deaf children ranged from 0% in Norway to 43% in Cuba (9% United Kingdom, 18% Canada, and 35% United States). Those who said they would offer this tended to use the "autonomy" argument—that if this is what the parents chose, and they were able to make a fully informed autonomous decision, then it would be acceptable to offer the technology to them.

Opinions of the British Deaf Association

The British Deaf Association (BDA) is "the UK's largest national organization run by Deaf people for Deaf people" (BDA website). The BDA has written a policy on genetics (updated in May 2003) that does not comment on prenatal genetic diagnosis for deafness with the intention of having deaf children (i.e., selective abortion for a hearing fetus). However, it does stress concern over the use of prenatal genetic testing with the selective termination of "deaf" pregnancies. In addition, they *demand* that

> all genetic counsellors should receive Deaf awareness training to ensure a clear understanding of the Deaf community and Deaf culture . . . [and that] . . . parents are not formally or informally pressured to take prenatal tests or to undergo termination where it is discovered that the foetus is deaf. (BDA 2003)

Therefore, the BDA advocates choice and informed decision making. They believe that this is not currently in place in the United Kingdom, and in order to rectify this, they want geneticists and genetic counselors to all potential parents of deaf children to present the Deaf culture perspective. With this scenario, hearing parents would receive information about Deaf culture and the community to which deaf people can belong prior to choosing to have prenatal genetic diagnosis and selective abortion for deafness.

CONCLUSIONS

Genetic testing for deafness is a very sensitive issue. Many within the genetics profession assumed that there would be a seamless integration of molecular genetic research into clinical practice and that genetic testing for deafness would be a positive issue for all concerned. There has been surprise and disbelief that not everyone could view this positively. Deaf people may feel devalued by the potential use of genetics (Middleton et al. 1998) and may feel threatened that genetics will in some way interfere with their culture. Since 90% of deaf children are born to hearing parents, the future of the Deaf community is in the hands of the hearing parent. This fact may not sit comfortably with many culturally Deaf people.

In looking to the future, it is clear that deaf people and the organisations that represent them should be involved in any policy decision making about genetics. There is always a greater need for more communication between deaf organisations and the medical profession. As more studies are done to look at how genetics services should be offered to deaf people and their families, a better clinical service will emerge. With regards to improving services for deaf people within the field of clinical genetics, there really is a collective accountability for this:

> while genetic counsellors have the responsibility to learn about the deaf community, . . . they cannot do this alone. There is a shared responsibility to educate each other. Deaf people need to be informed about genetics in order to be advocates for themselves. Geneticists need to be informed about deafness in order to perform their jobs in an ethical way. (Jordan 1991)

It is therefore vital that more research is done in this field and more public debate is initiated so that appropriate and effective services can be developed within the many agencies involved with deaf and hearing individuals and their families. The arguments for and against prenatal genetic diagnosis for deafness evolve with time. As it becomes possible to test for more non-life-threatening conditions, the boundaries of ethical practice will be pushed. Maybe it is time for us as a society to assess how we want to regulate this.

ACKNOWLEDGMENTS

The main body of work presented in this paper was published in the *Journal of Genetic Counseling* (Middleton, Hewison, and Mueller 2001). This work was originally done as part of a Ph.D. thesis. The author would therefore like to acknowledge and thank the Ph.D. supervisors, Jenny Hewison, and Bob Mueller, for their support and guidance.

REFERENCES

American Medical Association. Ethical issues related to prenatal genetic testing. Archives Family Medicine. 1994;3:633–42.

Anstey KW. Are attempts to have impaired children justifiable? Journal Medical Ethics. 2002;28:286–88.

Australasian Bioethics Information. Designer babies/go ahead to screen out deafness. Australasian Bioethics Information newsletter (serial on the Internet). 2002 Sept 27. Available from: http://www.australasian-bioethics.org/Newsletters/047-2002-09-27.html

British Deaf Association (homepage on the Internet). London: The Association. Genetics policy statement. Available from: http://www.britishdeaf-association.org.uk/policy/index.php?page=Genetics

Brunger JW, Murray GS, O'Riordan M, Matthews AL, Smith RJH, Robin NH. Parental attitudes toward genetic testing for pediatric deafness. American Journal Human Genetics. 2000;67:1621–25.

Cohen MM Jr, Gorlin RJ. Epidemiology, aetiology and genetic patterns. In: Gorlin RJ, Toriello HV, Cohen MM Jr, editors. Hereditary hearing loss and its syndromes. Oxford: Oxford University Press; 1995. p. 9–21.

Dolnick E. Deafness as culture. The Atlantic Monthly. 1993;272(3):37–53.

Fletcher JC. Deaf like us: The Duchesneau-McCullough case. L'Observatoire de la genetique—Cadrages. 2002 July/Aug;(5).

Infertility Treatment Authority. Policy in relation to the use of preimplantation genetic diagnosis for genetic testing. Departmental policy, Melbourne, Victoria, Australia. 2003.

Jordan IK. Ethical issues in the genetic study of deafness. Annals New York Academy Science. 1991;630:236–39.

Kelly J. Designer baby to have perfect hearing. Herald Sun. 2002 Sept 21.

Levy N. Deafness, culture and choice. Journal Medical Ethics. 2002;28(5):284–5.

Martinez A, Linden J, Schimmenti LA, Palmer CGS. Attitudes of the broader hearing, deaf and hard of hearing community toward genetic testing for deafness. Genetics in Medicine. 2003;5(2):106–12.

McLellan F. Controversy over deliberate conception of deaf child. The Lancet. 2002;359(9314):1315.

Meadow KP. Personality and social development of deaf persons. In: Bolton B, editor. Psychology of deafness for rehabilitation counselors. Baltimore: University Park Press; 1976. p. 67–80.

Middleton A. Attitudes of deaf and hearing individuals towards issues surrounding genetic testing for deafness (Ph.D. dissertation). Leeds (U.K.): University of Leeds; 1999.

Middleton A. Parent's attitudes towards genetic testing and the impact of deafness in the family. In: Stephens D, Jones L, editors. Genetic hearing impairment—Its impact. London: Whurr; 2004.

Middleton A, Hewison J, Mueller RF. Attitudes of deaf adults toward genetic testing for hereditary deafness. American Journal Human Genetics. 1998;63:1175–80.

Middleton A, Hewison J, Mueller, RF. Prenatal diagnosis for inherited deafness: What is the potential demand? Journal of Genetic Counseling. 2001;10(2):121–31.

Noble T. Embryos screened for deafness—a quiet first for Australia. The Sydney Morning Herald. 2003 July 11. Available from: http://www.smh.com.au/articles/2003/07/10/1057783286800.html

Padden C. The deaf community and the culture of deaf people. In: Wilcox S, editor. American deaf culture. Silver Spring, Md.: Linstok Press; 1980. p. 1–16.

Reardon W. Connexin 26 gene mutation and autosomal recessive deafness. The Lancet. 1998;351:383–4.

Savulescu J. Deaf lesbians, "designer disability," and the future of medicine. British Medical Journal. 2002;325:771–3.

Schein JD. Family life. In: At home among strangers. Washington, D.C.: Gallaudet University Press; 1989. p. 106–34.

Schum RL. Communication and social growth: A developmental model of social behavior in deaf children. Ear and Hearing. 1991;12(5):320–7.

Spriggs M. Lesbian couple create a child who is deaf like them. Journal of Medical Ethics. 2002;28:283.

Stern SJ, Arnos KS, Murrelle L, Oelrich K, Welch K, Nance WE, Pandya A. Attitudes of deaf and hard of hearing subjects towards genetic testing and prenatal diagnosis of hearing loss. Journal Medical Genetics. 2002;39:449–53.

Van Camp G, Smith RJH. Hereditary Hearing Loss (homepage on the Internet). Antwerpen: University of Antwerpen (cited 2004 July). Available from: http://www.uia.ac.be/dnalab/hhh/

Wertz D, Fletcher J. Ethics and genetics: in global perspective. 1999. Personal correspondence.

NEGOTIATING (GENETIC) DEAFNESS IN A BEDOUIN COMMUNITY

Shifra Kisch

In the winter of 1995, I became acquainted with the inhabitants of a Bedouin settlement in the Negev Desert of Israel. These people, whom I refer to as Abu-Shara or As-Sharat, after their common ancestor, Abu-Shara, are all of common descent.[1] This group, now in its seventh generation, numbers about 3,000 people, nearly 100 of whom are deaf. The use of an indigenous sign language is widespread among the Abu-Shara, and it provides the foundation of what I call an integrated signing community, shared by hearing and deaf people alike. Since my initial visit, I have spent three intensive anthropological fieldwork stints, a few months each, living with one of the families and engaging in participant observation. The first and longest stint was five months during the spring and summer of 1997.

Though deafness is common among the Abu-Shara, pursuit of cure or prevention of deafness is rare. The first time I heard of any such prescription was from Hakima, a woman in her fifties. Most of her children have already married, and yet caring for some of her young grandchildren is part of her daily domestic routine. Hakima's father-in-law was deaf, as are two of her grandchildren. Previous to her marriage more than thirty years ago, however, Hakima could already communicate quite fluently in the local sign language, as her mother's younger cowife was a deaf woman.

Hakima, lately complaining of headaches, asked me to accompany her to one of the elderly local healers. While on our way, Hakima told me that one of her daughters-in-law had been instructed by a healer in the further north Palestinian town of Dahariya not to breastfeed her newborn child. He suggested that otherwise the child might be deaf like some of his nephews. Hakima then complained that today's young women are so eager to waste money on milk substitutes that they find the strangest excuses.

Before we left the local healer's house, I asked the healer if she was ever asked for intervention regarding deafness. She looked at me with amazement. I turned to Hakima and answered the healer with a supposed example: Has she ever been visited by pregnant women who sought a solution for deafness? She answered in a derisive tone, as though she was being bothered with trivialities,

> I treat problems [as she spoke, she stiffened her face to demonstrate suffering] . . . deafness [her face brightened up] is from God.*

Another man from the Abu-Shara, who also makes amulets, said, in response to a similar question: "God creates a deaf person, gives him a good job, grants him with good looks, a good brain, there are hearing people who don't have any of this." In both cases, a contemptuous tone accompanied their words, and later on, both went on to proudly demonstrate that they only treat cases of real suffering.[2]

Hakima, like many other deaf and hearing members of this community, regularly encounters different attitudes, discourses, and practices in which the endeavors to heal, fix, and avoid the birthing of deaf individuals is assumed to be fundamental and crucial. The medicalization of deafness is being introduced into this community through two main factors: a genetic intervention program that is attempting to reduce the occurrence of deafness and special education programs for

---

*Square brackets ([ ]) indicate either descriptive elaborations of the interaction or my own comments and questions.

deaf children that promote audiometry and hearing aids. Medical discourse is translated and partially reproduced in local discourses but not fully embraced.

The exposure to medical discourse occurs in various settings, and genetics in particular has acquired a powerful grip on popular imagination (Nelkin and Lindee 1995). Medical and scientific discourse is commonly employed in the promotion of health and cosmetic commodities. Medical discourse is also introduced through high school education, as in lay encounters with various members of the dominant Jewish society. In such lay or semiprofessional encounters, the modernization theory and the utilization of modern state services such as health services are often introduced. When it comes to deafness, however, for most members of the community, medical discourse is most prominently conveyed through the actual genetic research conducted and the subsequent attempts to introduce an intervention program.

Several years ago, a genetic research team from the nearby district university hospital identified recessive, nonsyndromic, autosomal deafness among the Abu-Shara (Scott et. al 1995). The study used rapid screening tools to determine carrier status for this specific mutation (Scott et. al 1998). About 100 members of the As-Sharat participated in this research, providing genealogies and blood samples. This was followed by attempts to inform and induce more members to provide blood samples and develop a genetic database and consultation project to enable the disclosing of genetic compatibility to potential marriage partners without having to initiate a special mutual test, and without disclosing individual carrier status.[3] However, compliance was limited.

I did not follow any actual counseling sessions during my fieldwork for this study, but I did witness people's decision-making process in daily life and the way in which they acted and referred to medical discourse and medical reasoning. In light of the growing exposure to medical discourse, it seems that a complex negotiation process is being held as to the plausibility of the genetic explanation of deafness and its resulting implications. Direct or spontaneous discussion of the findings of the genetic study and its ramifications is infrequent and tends to be indirect and implied. Detection of a child's deafness or the presence of a deaf person do not provoke discussions about the origin of deafness. Deafness, to a large extent, is not perceived as something that calls for explanation. Thus, dealing with the question of deafness and its origins predominantly followed referrals to the medical discourse, or simply to the attribution of deafness to consanguineous marriage.

Before turning to these local accounts, I will provide some necessary context with two brief introductions. First, I will introduce the Bedouins residing in the Negev, the southern arid region of Israel. Second, I will concisely portray the nature of this integrated signing community. Following this general introduction, I will turn to the primary concern of this paper: the local accounts of deafness and what I consider to be their significant subtexts.

SOCIOPOLITICAL SETTING

The native Arab inhabitants of the Negev are former nomads commonly referred to as Bedouins. Since the establishment of the state of Israel, they have undergone drastic economic, social, and political upheavals that have resulted in deep destruction of the social organization. The Israeli government made a concerted effort to settle the Bedouins in a way that would minimize the use of land resources through the establishment of urban settlements. At present, seven such townships exist, gathering nearly half of the Bedouin population living in the Negev, which is estimated to be more than 120,000 people. The remaining Bedouin population is deprived of basic infrastructures (roads, water, electricity, sewage), as they inhabit settlements formally unrecognized by the state. The community discussed in this paper resides in such a settlement.

The Bedouins constitute 25 percent of the population of the Negev subdistrict and live separately from its Jewish population. However, the lack of economic activities in both the unrecognised villages and the townships as well as the scarce qualified manpower render the population dependent economically on the Jewish sector and very sensitive to economic recessions. Most men, but only a few women, are enrolled in the Israeli labor market; unemployment rates are among the highest of the country. Within contemporary Bedouin society, there are considerable variations in education, lifestyle, and women's status. Among the Abu-Shara, illiteracy rates are still high among women, as well as among elder men. Among the younger men, though, several have in recent years attained an academic professional education and studied law, pharmacology, or medicine abroad. Sixty percent of the

Bedouin population is under the age of sixteen, fertility rates are among the highest in the country, and consanguinity and polygamy are common. All Bedouins, as citizens of Israel, obtain a national health insurance and fall under the compulsory schooling law.

Access obstacles, state priorities, and language barriers limit actual exercise and use of these rights and services, however. The government implements policy that is perceived to meet state interests (such as security, population control, and controlling welfare expenses), but other affairs suffer neglect. A case in point is the state's interest in reducing the rate of children who are born deaf, or for that matter, of children that require special services. State authorities have invested considerable funds in a scheme to reduce congenital diseases among the Bedouin population, and part of the genetic intervention program mentioned here is financed by such funds. At the same time, little if anything is done to improve the poor living conditions among the Bedouins.[4] Supreme court intervention has been required to force the state to provide some basic rights and services, such as instructing state authorities to establish health clinics in Bedouin villages and to connect schools there to the water system.[5]

AN INTEGRATED SIGNING COMMUNITY

In this paper, I refer to the Abu-Shara as an "integrated signing community," but I recognize that *community* is a problematic and frequently criticized term. Here I am using it only in the restricted sense of a linguistic speech community.[6] The widespread use of an indigenous Abu-Shara Sign Language (ASSL), shared by hearing and deaf people alike, is the foundation of this integrated signing community. The anthropological and sociolinguistics literature contains evidence for few other such integrated signing communities (Branson et al. 1996; Johnson 1991; Groce 1985). The best-known research is Groce's study of the Martha's Vineyard community. I use the term here to distinguish the Abu-Shara from other signing communities that consist mainly of Deaf people.

Harlan Lane et al. (2000) have argued, in their historical comparison of Deaf communities in New England in the early-nineteenth century, that beyond the prevalence of deafness, different genetic patterning contributes to the spread of sign language into the hearing environment. Given the recessive pattern among the Abu-Shara, deafness is indeed widely distributed among families in the kin group.[7] Social realities are not determined by genetic patterns. However, genetic patterning may contribute to the blending of deaf and hearing people. Among the Abu-Shara, most hearing people are familiar with deaf individuals; many have a first- or second-degree deaf relative. All hearing people are exposed to the viability of fluent communication in sign language and to assimilated, well-integrated deaf figures in the community. Most hearing As-Sharat are familiar with the local sign language, though with varying degrees of fluency. This implies not only ease of communication within the community, but a perception of deafness as merely a condition calling for use of signed language rather than spoken communication.

I wish to underscore the distinctiveness of this social reality: deafness does not produce social marginality or isolation, nor does it serve as a postulate for social alignment as a distinct social group. There is no distinct Deaf community in spite of abundant deaf people and the existence of a local sign language. All married deaf people have hearing spouses. No social events roles or activities are reserved for deaf people due to their deafness, except within the education system, which has brought about separation between deaf and hearing.

I have argued elsewhere (Kisch 2000; Kisch n.d.) that deaf members' status, like that of hearing members', should be seen in light of a given dialectical communication web. This web is comprised of multiple languages (Hebrew, Arabic, ASSL, Israeli Sign Language); language modes (signed, spoken, written); and language domains representing discourses not shared by all (gendered discourse, medical discourse, religious discourse, and others). Many of these need to be translated and mediated to various members of the community. Members of the community are located at diverse overlaps of these communication domains and are therefore in a position to translate or mediate for different people in different situations; yet in other situations, they would rely on other's mediation and translation. Thus I have argued that the reliance on translation and mediation is common practice and not restricted to deaf people.

This background is crucial to understanding the context in which the subsequent local accounts of deafness take place. The communication web is permeable and dynamic, enabling the infiltration of stigmatizing narratives that may marginalize and exacerbate the position of deaf people. These processes most evidently occur through encounters with the medical and education establishment. Interactions with the medical and educational establishment can be depicted by what Talal Asad (1986) called "the inequalities of languages," referring to the permeability of respective languages in relation to the dominant forms of discourse. While Asad uses this concept to address the constraint of "cultural translation" and ethnographic work, I find the concept useful in a broader context. It is relevant for the encounter between a linguistic minority and the dominant establishment, as well as to the authority of medical discourse versus local lay discourse.

TRANSLATING MEDICAL DISCOURSE

Medical discourse is anchored in scientific discourse. Scientific discourse presumes to constitute a representation of the scientific object, that is, laboratory facts without mediation or human agents (Latour 1993). Local discourse, however, refers to medical discourse as a supplier of one explanation among many, and thus emphasizes the involvement of mediation, often explicitly referring to the "explanation of the doctors."* This implies that one could question the exclusiveness of the explanation, negotiate it, and examine it according to its agents, prescriptions, and implications.

It seems that even a speaker who refers to the medical explanation as a definite, all-embracing explanation does not necessarily grant it precedence. Hence, it is not the "knowledge" of the medical explanation but its positioning in the local discourse that raises important questions concerning social preferences. Every account presents different paradoxes among various social considerations. The paradoxes and the attempts to settle them are the background for offering the local accounts. The examples in the following section provide the background for the local accounts that follow and release them from the narrow question of knowledge and ignorance.

"The Doctors Say"

Exposure to medical discourse occurs in diverse settings. For most members of the community, however, medical discourse related to deafness

---

*I therefore use the terms *medical discourse* and *medical explanation* alternately throughout this essay.

was most prominently conveyed through the actual genetic research conducted in the community and subsequent attempts to introduce an intervention program. Most members of the community are aware of the existence of the genetic claims, even though compliance to the genetic intervention program was low, and even though they might not fully comprehend the concept of recessive hereditary traits or statistical probabilities. It became common knowledge that the medical view denounced consanguineous marriages, but it was not always clear who was regarded as a "relative" by the medical view. The following cases demonstrate the clear presence of the medical explanation in the local discourse and how members of the community selectively refer to the medical discourse.

The following account was given by Abu-Saalem, a man in his late fifties, who married two women, both his cousins. All of his children are hearing; one of his many grandchildren is deaf.

> Today one hears from the doctors that if you marry from out-
> side the family, you will not deliver mute [deaf children] . . .
> I don't think that this has anything to do with family or blood.
> . . . A matter of retardation, that I do believe, I think it's from
> the blood. Today they [the doctors] can check if one is suitable
> for the other or not; it's a new thing! If it is the same blood, a
> retarded child will come. I can believe that. But mute [deaf]—
> this isn't connected to this thing, I'm sure . . . I don't believe
> this thing.

Abu-Saalem is clearly familiar with the genetic claims; he is even familiar with the genetic testing. Yet with regard to deafness, he chose to phrase the doctors' position simply as a recommendation to refrain from consanguineous marriage. In other words, his initial account reduces what the doctors state to terms of marrying within or outside the family.

Most local references to the medical explanation treat it as an explanation that ultimately denounces consanguineous marriages. This is so in spite of the fact that, ironically, the applied model, as proposed, was intended to emphasize that the As-Sharat could marry each other while still minimizing the chances of bearing deaf children. Testing couples for genetic compatibility would accomplish this.

The medical explanation is present in the community, but it has traveled a long way from the laboratory and the medical discourse that serves it, through the language of genetic counseling, to the way it is used in the local discourse. In a study that focused on nonexperts' definitions and perceptions of the risk values presented by the genetic counselors and the medical team, it was argued that a lot of the information regarding the risk is lost in translation. Thus, likelihood was translated into absolute theoretical categories, and risk was converted to prohibitions and clear prescriptions (Parsons and Atkinson 1992).

Among the As-Sharat, the same individual may translate the medical discourse in different ways, depending on context. I find Rapp's (1988, 1999) notion of negotiating most appropriate to understanding these seeming inconsistencies.[8] Rapp contends that genetic counselors themselves deal with the translation of medical discourse into popular and more usable language. In this process, many concepts are under negotiation, albeit not formally. Not only is medical knowledge constructed, but power relations and popular social knowledge are also established.

While some reproduce the medical discourse selectively, others reproduce it quite extensively but restrict it to certain situations while rendering it irrelevant to others. One example of such a rendering was provided by Ikhsen. I met Ikhsen in the shack that served as the local medical clinic. He noticed me talking to a deaf woman, and after he inquired about me with one of his sisters-in-law, who was also present, he turned to me. Because he insisted on staying in the shack, most of the women went outside to wait. His sister-in-law stayed inside.

> Ikhsen: We have all become deaf at Abu-Shara, there are a lot of
>   deaf people. Every family you will encounter has two or
>   three, and there are also a lot of heart diseases in every
>   family. We have a long list of diseases! . . . Deafness is
>   not a disease; it's from marriage. . . . If our sons go and
>   marry outside, far away, it will be much better.
> Interviewer: And what about the daughters?
> Ikhsen: Our daughters? The elderly will marry them.

He smiled when he said this. I looked at his sister-in-law, who sat beside me. She sighed and raised her hands. He had said the last sentences in Hebrew, and she did not understand them. Then Ikhsen continued:

Ikhsen:	They want . . . do you think it is better for a woman to go and get married far away?! She doesn't know who these foreign in-laws might be. It is better for the women to marry nearby!
Interviewer:	[But didn't you say that deafness is from marrying relatives? I asked.]
Ikhsen:	Yes, this is a problem. But it's still better to marry here.

Marriage outside the family has long been accepted for men and could even be a source of prestige for them, but for women, the contrary is true. They are expected to marry inside the family. Men often adopt prescriptions to refrain from consanguinity themselves, and at the same time, adhere to consanguinity as far as their daughters and sisters are concerned. Men regularly use medical discourse in reproducing patriarchal principles, according to which the introduction of brides from other social groups is desired.[9] However, handing over daughters as brides to foreigners can be undesirable. When nonreciprocal social relations are involved, the receiving of brides—from a group that one would not marry his daughters to—can be understood as an act of domination. The hypergamous principle sustains an endogamous gap between men and women. Thus, elite groups have high rates of endogamy among women, although men take brides from other groups (Kressel 1992).

When referring to the past, people often suggest that there was a lack of alternatives to consanguinity, even if its consequences were known. As the following account illustrates, this incongruity is not a new one. Fawaz is a rather affluent man in his late forties with a number of deaf siblings. He has two wives, both of Abu-Shara. Fawaz told me about his elder father's consultation with doctor Ben-Assa many years ago.[10]

Abu-Assa, after research, suggested that if they wanted to get rid of all this phenomenon [deafness]—just stop marrying one another, and there will be no more deaf-mutes. At that time, if you say something like that to the elderly—owwa! Abu-Assa examined this thoroughly; this thing comes from the blood not being compatible. He told them! They listened and knew, but they didn't have a choice . . . they were interested, but there was no one to give them [brides].

The argument put forward here, that consanguinity is a last resort, is a common one. It relates to power relations, class, and prestige within the Bedouin society. Higher rates of kin endogamy are common among Bedouin who lack claim to noble origin (Marx 1967).[11] Fawaz hinted at the existence of class differences in the past. Others indicate that they exist today as well. However, it appears that Fawaz was too proud to enhance this image, and therefore, he added, with a smile,

> I will tell you the reason exactly; they know that there is a 90 percent chance that their son will be mute [as a result of consanguinity]. The problem—we have pretty girls—it hurts to give them away.

The reproduction and representation of the medical discourse has yet another discursive value, that is, its contribution to one's presentation of self. In a sociopolitical setting strongly dominated by popular policy concepts of "tradition" and "modernity," people are often conscious of the binary division that regards the acknowledgment of scientific authority as a basic condition of being modern. To demonstrate one's familiarity with the medical discourse is therefore a way to associate one's self with modernity. I do not wish to imply that it is therefore a superficial practice; rather, embracing the biomedical discourse and the concept of the modern are among the multiple and often conflicting internalized idioms and positions of this community. I then wish to join Abu-Lughod's reservation from what can be perceived as Goffmanesque "empty acts of impression management" (1986, 238).

Speculating

The medical establishment's model presupposes that the high rate of deafness necessitates intervention in one of the most sensitive social arenas, marriage arrangement. My research suggests that the local accounts of deafness are offered mainly as a result of concern with the implications of the genetic explanation. Deafness is commonly perceived as manageable, but finding suitable marriage partners (for deaf or for hearing individuals) is a source of growing concern. From this point of view, it is possible to view the local accounts as alternatives to the medical explanation.[12]

Local accounts draw on various existing idioms related to procreation. The attitude towards the origin of deafness is mostly speculative; most people I spoke to raised a variety of possibilities and did not necessarily endorse one particular stance. Offering one explanation usually led to it being challenged by another. As a result, a new version would be offered, and so on. This dialectic situation makes it difficult to present the different accounts for deafness as an organized typology. Hence it would be most constructive to observe how people frame their speculations.

Abu-Musa is in his fifties. He has two related wives and twelve children. Several of his siblings are among the elderly deaf people in the community. He said the following things to me while helping depict Abu-Shara genealogies.

One of the sons of our ancestor Abu-Shara [the ancestor of the whole kin group] took a wife from the Shrukh [a different family from another region]. My father married one of her daughters, and from this union came three mute and three speaking. Then it spread. [Through their children? I asked] No, not at all.

I think deafness isn't a disease; it is God's way. It's being said it comes from kin-marriages, but I do not believe." [So what do you think? I ask] "When one plants seeds, one watermelon comes out like this and another like that. [With his hands, he demonstrated a round one and an oval one. And this one is not as good as the other? I ask.] It's a very good one; a watermelon did come out, didn't it?

Today the doctors say its bad to marry in the family. You must mix, they say. I see some who married not at all from their relatives, and they got much worse. If only the child was mute! But this child is retarded. There is much worse than muteness; the mutes manage; they manage very well. But the retarded, there is no question of managing. There is a difference between hearing and not hearing, I say it, but I don't believe it's from the family. You know, personally, I don't believe as the doctors say; [he raises his voice determined] "yes!" I say; maybe it's not certain.

The words of Abu-Musa underscore a few general attributes of the local accounts of deafness. As mentioned earlier, it is significant to observe the presence of the medical claim in local discourse. It is also significant to note that Abu-Musa left room for multicausality. This in itself challenges the genetic explanation—which is exclusive—and makes it negotiable.

Abu-Musa employed several challenging possibilities; each one of them refers to existing local notions related to procreation. It is possible to highlight these notions, recognize their subtext with regard to the medical discourse, and follow how they give rise to yet another translation or elaboration drawing on medical discourse and yet other local idioms.

Abu-Musa started by suggesting that deafness was "brought into" the family by brides from outside the Abu-Shara family. The subtext suggests a subversion of the medical discourse on inbreeding. While medical discourse denounced consanguinity, Abu-Musa referred to the patriarchal notion that related women, belonging to the same kin group, carry the familiar, acclaimed family traits. In contrast, unknown women carry unknown traits. As I will further clarify, this account entails some negative connotations about deafness and women.

Abu-Musa then suggested that deafness reflected a difference in seeds. This is a common patriarchal metaphor, often referred to as "the seed and the soil" (Delaney 1991; Leach 1969). In sharp contrast to the previous notion of "brides from outside," the idea of "soil" deprives women of any significant contribution in determining the major character of their progeny. Deafness is then a patrilineal trait (represented in the seed). The idea links this natural metaphor from agricultural reproduction to God's way, which does not require explanation or intervention, but acceptance. This led Abu-Musa to express his doubts as to the extent to which deafness is a problem at all.

Despite the occurrence of the accounts depicted above, "min Allah" (from God) is by far the most prevalent response to dealing with the origins of deafness. In fact, this response does not constitute an explanation of the source of deafness, but rather draws attention to God's mysterious ways. It does however have a subtext.[13] The will or way of God is an expression for that which is beyond investigation or change (Comaroff 1981). Referral to "God's will" sees the situation as "natural," as no more than a variation that needs no explanation.[14]

This argument implies that causality is not relevant. Perhaps the search for explanations should not even be attempted, for the will of God is beyond human comprehension. Such comments often act as a closing remark at the end of a long and intricate conversation on the origin of deafness, or as a means for cutting it off. It suggests that the answer or source of the phenomenon should not be sought in human action.

"Brides from Outside"

Let us have a closer look at Abu-Musa's first account: deafness was introduced by "brides from outside." This is the only account that implies a negative association with deafness. It suggests that deafness is not an accepted family feature but rather a tainted element from the "outside," as it "blames" foreign women. While this characterization can be seen as compatible with the medical discourse, in its attempt to present deafness as an undesirable contaminating element, with regard to marriage arrangements, it implies the opposite. The genetic project points to consanguinity as the source of problems, but the foreign brides account suggests that "mixing" with foreign families might be a source of unknown and possibly undesirable traits.

Pointing to "brides from outside" as accountable for deafness is akin to the medical discourse that describes deafness as undesirable. It stigmatizes deafness, but it also stigmatizes women as carriers of undesirable foreign traits. It is therefore not surprising that women in their alternative speculations would question both of these notions. Furthermore, women have their own reasons for preferring to marry within their familiar kin group—because that is where their social network can be put into action and serve as a source of social power.[15]

Samira is a woman in her forties who is married to her paternal cousin. She demonstrates the advantages of being married to a relative: their household is surrounded by those of her brothers and her mother, and she is a cherished sister, authoritative and full of humor. Samira dismissed the idea that the foreign origin of her maternal grandmother would be the reason for her mother's and brother's children's deafness.

A man will never take responsibility upon himself, never say
about himself that he has flaws; and a man has many flaws.
[Samira laughed and continued.] Yes, that old woman was from
the Shrukh [family], but the Shrukh, they don't have mute
people at all! [So where is this story from? I asked.] I don't
know; it's a mistake . . . but there is no mistake in deafness, it
is from God.

Samira was familiar with the doctors' approach; her response to
the compatibility tests indicated she was far from ignorant regarding the
possibility of having such tests performed. "I know the doctors say in
our family there are many problems because people marry each other.
But look. If everyone does the test, they'll stop marrying one another
altogether! There are people who married from outside—but the child
still has problems." Samira did not discard this idiom on the whole, but
she rejected its association with deafness.

Inheritance, Blood, Seeds

Like Abu-Musa, Samira wished to relativate the cruciality of deafness.
Like Abu-Musa (who used the watermelon metaphor), Samira too
referred to God and nature.

It's from God . . . For example, in agriculture, the root sprouts
into seven generations. Here, take us; my mother is mute, and
I don't bring any deaf children. Perhaps my daughter will not,
perhaps my daughters' daughters will; perhaps my sons' sons will;
it is present in the inheritance. It passes on in the blood; it is
mostly present in the man. You plant a watermelon, a watermelon comes out; you plant a melon, a melon comes out. What
you plant is what turns up. . . . It is from the man, not from
the woman.

Images from the plant realm are used to illuminate that this is a "natural" phenomenon. Indeed, as Mary Douglas points out, "natural symbols" have the power to camouflage the cultural as the natural, which
is not subject to human act (Douglas 1996). According to the domi-

nant metaphor of "the seed (semen) and the soil (womb)," women are connected to the "natural" world (Delaney 1991; Ortner 1974). A woman nurtures and supports (pregnancy, breast feeding), but she does not determine the newborn's essence. As Samira clearly suggests, if one would plant a melon seed in different soil, one still would not get a watermelon.

Samira summarized this perspective: "The woman doesn't place anything! She doesn't do anything, doesn't implant and doesn't uproot." All the women burst into laughter. Samira's deaf mother was also present, and until that point, one of the young women translated to sign. But Samira obviously wanted the stage for herself. At that point, she was at the peak of her enthusiasm and was speaking and signing dramatically. Later on, in response to other comments made during the conversation, she relented a bit: "Most is from the man, not from the woman. . . . Maybe 1 percent from the woman and 50 percent from the man."

Samira was negotiating, sometimes even negotiating in numbers. However, the consistency of her words does not lie in presenting a seamless theory, but in the discursive value of her arguments. In that respect, her message is consistent, and the women's laughter implied that they clearly recognized the meaning: "Don't burden women with all this."

The underlying concepts employed by Samira, Abu-Musa, and others should be highlighted: inheritance, blood, and seeds (the Arabic word means both seeds and semen) cannot be understood without their patriarchal context.[16] These terms do not refer symmetrically to men's and women's procreative contribution in determining their children's blood. Some accounts may appear as local versions of the medical account, but they often carry a quite contrary subtext in which matrimonial pairing is rendered irrelevant.

Referral to blood emphasizes men's role in procreation. Thus, it enables a challenge to the medical-genetic assumption that interference is needed in marital pairing. The term *blood* often refers to social relations more than to an individual's body.[17] Unlike in genetics' reference to blood tests representing the genetic makeup, the term *blood*, as locally used, represents one's patrilineage. It is irreplaceable and not related to marital pairing, and as such challenges the possibility of preventive action. Bourdieu (1977) demonstrates the possible variation in male and female reading of kinship relationships. These cultural idioms can indeed be read and employed in various ways. They are gender dependent, and men and women use them differently with regard to the context and subtext of such readings.

"Khulf"

The following ethnographic example demonstrates yet another account of deafness. Furthermore, it calls attention to the fact that not all wish to propagate endogamy or to praise the family's endemic traits. Many, wishing to conform to modern discourse and its rejection of what is perceived to be "traditional," condemn consanguinity. Yet it is not just the elderly who combine various explanations and draw on local theories of procreation to negotiate the genetic implications.

Faher is an educated and rather religious young man, married to his second cousin, and father of several hearing children. Obviously, he has incorporated major aspects of the genetic discourse.

> We have many healthy children; they are OK. But if you will check their head, bones, heart, you will find many shortcomings. Why did God choose to do this to us? That is helpless talk. Will we blame and burden Allah for everything bad? [These things were said in the presence of both his parents, who were expressing their scepticism. He impatiently turned to his father.]
>
> That's why they ask at the hospital: Your father, does he have anything? Sugar [diabetes]? Heart? That is the way it goes; hereditary, from one generation to the other! [I then asked Faher again regarding deafness: So that is what you think it is?]
>
> When it comes to deafness there must be something else. [Why? I asked.] I know people who don't have deafness in the family at all, and it still happens. There is something in heredity, but the first reason for deafness is different. When a pregnant woman sees something, she gets so engaged with it, that almost the same element is marked in her child. [Faher here referred to yet another local idiom, a form of marking, locally referred to as "Khulf," which is often mentioned in relation to deafness and other conditions.]

For example, the first daughter of a young couple I knew had fair hair, despite the fact that both of her parents had dark curly hair. I was acquainted with her mother well before she married. I used to visit her often while she was pregnant, and I joined her at the prenatal care clinic. On one of my later visits, her sisters told me, while she brushed the toddler's hair: "She [the mother] was keen on you and brought like you. She marked through you."

The mother was perceived as the active agent who marked the child she carried, though the marking was not intentional. It is noteworthy however that it was not the person who evoked the mother's attention that imposed his or her traits on the newborn. This is an important distinction. If the beggar, the fair-haired person, or the deaf signer were perceived as the active party (as the people who imposed their characteristics on the newborn), it would suggest that they possessed extraordinary or even witchery powers, and thus the marking would be unavoidably stigmatizing. Though compassion was also mentioned with regard to marking, different things can be perceived as occupying one's attention about deafness, such as some other prominent feature of the person. A woman once illustrated Khulf by suggesting that she might have marked her son with deafness following her admiration for a neighbor who was a strict and sturdy educator. He was deaf, and she used to observe him while pregnant.

What then is the subtext or discursive value of this account? Unlike genetic explanations, marking is typically monogeneric, that is, it has nothing to do with marital or procreative pairing. It is related to what is perceived to be female nature, challenging the idea that anything can or should be done about it. Like other accounts, it is a speculation drawing on an existing idiom, and it is not presented as an all-embracing explanation of people's deafness.

Faher even extended Khulf to a broader theory on deafness. He emphasized that it was not deafness in itself that evoked marking, but it was the visibility and appeal of signing. In this way, he attempted to offer an alternative explanation for how a child can be born deaf, and he suggested what characteristic about deafness could draw the attention of a pregnant woman. Faher also applied the Khulf idiom to provide an account of the high prevalence of deafness among the Abu-Shara, he thus concludes "We live close together, signing is everywhere, and signs engage one's attention."

Stigma

It is possible to examine two aspects of each local explanation of deafness. One refers to the meaning an account attributes to deafness, that is, what can be learned about perceptions of deafness from a certain way of explaining it. The other aspect relates to the offering of an explanation as practice and its role in the local discourse. The first aspect is secondary to this paper, which focused on the latter. However, although attitudes towards deafness are complex and even ambivalent, local accounts reveal a relative lack of social labelling.

It is significant that there is no single local myth as to the origin of deafness—neither of a particular individual's deafness nor of the high occurrence of deafness. None of the idioms (whether "brides from outside," "seeds," "marking," or others) is reserved to explain deafness only. These modes of explanation also are used to account for other differences between members of the family (height, hair color, character, or general resemblance).

To clarify this last point, it might be useful to resort to a comparative example. I once visited a young woman, Hannan, from a neighboring kin group where deafness had not occurred before she was born. The difference between her account and those of the Abu-Shara women was striking. She kept on referring to her deafness in the following terms:

> It's difficult, very difficult. What can I do? I did not do anything wrong. My father, his heart is in pain over me. Where doesn't he drag me to? He begged that I be cured, and there is no one to help. [Her mother said,] Look at her, like a flower, pretty and doesn't hear and doesn't speak, poor thing. . . . May God watch over her . . . . Lord of mercy.

Among Hannan's kin group, deafness calls for a cure. Moreover, it seemingly needs to be confirmed that she is intelligent and has other good qualities. The tendency to attribute a wide range of faults due to one human feature is one of the characteristics of social labelling

(Goffman 1963). Hannan's deafness was explained by a specified event of maternal fright, a negative form of marking.[18]

> I am this way from my mother's belly. My mother remembers that when she was pregnant, she once woke up in fright from an airplane. . . . She too cried over me. . . . There was this noise like an explosion, and she felt something wrong.

From my acquaintance with Hannan and her family members, as well as with other Bedouins around the Negev, I learned of the common perception of deaf people as handicapped and ill-fated. This contrasted sharply with the absence of similar accounts among the As-Sharat, hearing or deaf. Deaf people among the Abu-Shara do not offer personal explanations for their deafness or for the deafness of others. However deaf members of the community, as others, have their own situated perspective with regard to deafness, marriage, and the social priorities entailed.

I visited Hannan with her former classmate Sabriyah. Sabriyah, herself deaf, was still unmarried at that time, but all three of her deaf sisters and her deaf brother had long been married. When Sabriyah declined two different offers to marry a foreign man (not from the Abu-Shara family), she told me,

> And if I marry this man [from a village in the south of Mount Hebron], who will I be there? How will they treat me? Where will I go when there are problems? Here we know each other; and if I have problems, I am close by. If he is rude to me, I will go to my neighboring sister. If he decides that my daughter should stop going to school, I will make a commotion; I will turn to everyone and not let this happen, whether she is deaf or not! [A year later, she married Said Abu-Shara , then fifty years old, becoming his second wife.]

Sabriyah's situation raises the question of whether, and to what extent, a genetic characteristic will be permitted to override social factors in a woman's preferences of a potential marriage partner, and whether it will determine her identity and those of her future children. As a deaf woman in Abu-Shara, Sabriyah has social resources that are based on a dense and accessible network. She perceives herself as an active and resourceful agent in the future of her children, deaf or hearing. Later in life, the future of her children and their status will influence her own status. What would Sabriyah's status be, and what social resources would be available to her as a foreign, deaf bride, if she married a stranger? How would his family members communicate with her? How would she and her children be treated, and what behaviors or characteristics would be attributed to her deafness? It appears that Sabriyah, like other women, even when aware of genetic risk, were not interested in the social risk required to find the answers to these questions.

CONCLUSION

I wish to conclude by returning to my initial remark on the multiplicity (and often seemingly contradictory) causality often present in local accounts of deafness and what might seem as incredible inconsistency. Faher initially criticized his parents for not understanding the simple principle of genetic inheritance. Willing to define the situation as a problem, he also argued strongly that heredity alone does not explain deafness. Faher altered what he perceived as strictly modern concepts with ones that might be seen as strikingly exotic. Abu-Saalem was fascinated with the ability of doctors to check people's blood for congenital defects like retardation, but maintained this could not be relevant to deafness. Ikhsen related consanguinity to both deafness and heart disease. He resorted to genetic discourse to support the patriarchal practice of importing brides, but genetic considerations seemed to lose their relevance when it came to local women. At first glance, the use of different, sometimes contradictory, explanations by the same person may seem inconsistent. However, this is so only when an attempt is being made at drawing cultural cosmologies of conception. Rather than try and solve these contradictions in an effort to formulate a unified theory of procreation, I have tried to demonstrate the way these explanations function in particular circumstances. In other words, I examine discourse as a practice and not merely as a key to cultural representations. Language and reasoning are not merely reflections of beliefs or ideas; they are modes of action. I do not wish to argue that Bedouins are more inconsistent than others. Rather, my working assumption was that social life and cultures are not consistent. Knowledge is situated, and people's statements in particular are highly situational. The consistency of the various accounts lies in their discursive value or subtext.

A range of deliberations mask or play down genetic considerations in matchmaking, and local speculations provide valuable insights into the range of competing considerations. People extract their explanations from the rich pool of perceptions and explanation patterns to which they are exposed. People modify the use of several explanations anchored in different dimensions of social life and discourse, and they sometimes choose conflicting ones. They make creative use of them in different contexts, while negotiating their contents. As a result of this, new subjects and contexts emerge. The discursive value of the explanations enables constant changing of the thematic frame in a particular conversation.

Medical discourse is translated and partially reproduced in local discourses but not fully embraced. Referrals to medical arguments or medical authority are highly situational and contextual. The genetic medical discourse is also subject to discursive interpretation, and therefore it does not easily impress its logic on local discourses.

The genetic explanation and prescription on procreation is partly represented in Abu-Shara, but people in the community consider many factors other than the avoidance of one specific genetic risk in seeking a marriage partner.

NOTES

1. To conceal people's identity, personal details might be slightly altered, and pseudonyms are used for all individuals, as for the kin group as a whole.

2. Seemingly, it can be argued that the scornful tone of the local healers does not stem from the fact that deafness is not defined as a sickness or as a problem, but from their experience of the situation as unsolvable, and their avoidance of failure. However, nineteenth-century European history saw several stubborn attempts to cure deafness, among them surgical. Although none proved to be effective, their practice persisted for a long time (Branson and Miller 2002; Volkov 1998).

3. An individual carrier detection program could more easily stigmatize women for "genetic incompatibility" and severely hamper their marriage prospects.

4. With regard to the unrecognized settlement, one could even claim the contrary—some of these conditions are manipulated to advance the imple-

mentation of state policy to reduce land use and negotiate land claims (Abu-Saad 1997).

5. A series of petitions have led the Supreme Court to reaffirm that the establishment of health clinics in Bedouin settlements is an essential service that the government must provide (Israeli High Court of Justice 2000, verdict 4540). In 1998, the Supreme Court's intervention was needed to instruct the ministry of education to connect all Bedouin schools in unrecognized settlements to the water system at once. The ministry argued in its defense that the settlements are temporary and are to be replaced by permanent ones. The school at the unrecognized settlement discussed here has been operating for more than thirty years.

6. A speech community is defined by Hymes as "A community sharing knowledge of rules for the conduct and interpretation of speech. Such sharing comprises knowledge of at least one form of speech, and knowledge also of its patterns of use" (Hymes 1974, 51). Hymes's definition is tinted by the referral to spoken language to indicate the actual use of language in interaction. The term does, however, apply to signed language as well.

7. Only four deaf parents among the Abu-Shara have deaf children; all but one have hearing partners and both deaf and hearing children. Two of these deaf parents create a generational depth of three successive generations of deafness; a deaf grandfather has deaf and hearing children, and one of his deaf sons has deaf children as well. In Desa Kolok, such generational continuity is more common (Branson et al. 1996).

8. See also Jansen (2000) discussion on the manner by which certain cultural idioms related to procreation can be employed to negotiate gender relations and other aspects of the changing social relations.

9. Such a marriage is not looked down upon and can serve to establish hierarchic relations or alliances between individuals and groups.

10. Dr. Binyamin Ben-Assa (1917–1976) served between the years 1954 and 1972 as government physician for the medical service for the Bedouin. He made a significant contribution to the understanding and treatment of morbidity among the Bedouin population in the Negev.

11. Such rates might be an indication of communal social isolation (Kressel 1992).

12. The academic discourse on "explanations" (and definitely the discourse on "belief") in itself might reflect a series of assumptions, which are related to modern science. Thus, these terms might grant authorization to the

medical discourse (Good 1994). Due to the lack of a more suitable term, I have used the term *accounts*, despite the problems involved in its use.

13. The attributing of phenomena to God's will is prevalent in another way, which is common in the Christian, puritan world (Groce 1985) but is not depicted in this instance. The will of God as an explicit expression of punishment or warning is a meaning that attributes moral and ethical contexts to the phenomenon.

14. In a study conducted in India, a process is described that leads deaf people and their families to accept deafness as "Prakritic"—a religious Hindu perception of "naturalness," which relates to the natural in a sense of belong-ingness to the world order. Therefore, this perception also releases believers from feelings of guilt and searching for remedies (Jepson 1991).

15. The structural, material, and ideological analysis of kin-endogamy is the subject of a classical and extensive anthropological debate (Abu-Lughod 1986; Bourdieu 1977; Kressel 1992; Khuri 1970).

16. The term *Wirathat Dam* represents a patrilineal heir who receives his inheritance due to blood relatedness. Damaya, the diminutive of blood (Dam), "suggests that maternal links are recognized, yet as a weaker form of kinship" (Abu-Lughod 1986, 51).

17. See Abu-Lughod (1986).

18. According to Groce, maternal fright was one of the most common explanations in the nineteenth century for all sorts of congenital defects, which were caused by panic or anxieties of the mother during pregnancy. In 1863, an article by Hawkins stated that nine out of every ten cases of congenital deafness emanate from the mother's anxieties. He even suggests that sequen-tial births of deaf children are caused by the mother's fear of giving birth to another deaf child (Groce 1985, 119–120).

REFERENCES

Abu-Lughod, L. 1986. *Veiled sentiments: Honor and poetry in a Bedouin society.* Berkeley: University of California Press.

Abu-Saad, I. 1997. The education of Israel's Negev Bedouin: Background and prospects. *Israel Studies* 2 (2): 21–39.

Asad, T. 1986. The concept of cultural translation. In *Writing culture: The poetics and politics of ethnography*, ed. C. James and G. Marcus, 141–64. Berkeley: University of California Press.

Branson, J., D. Miller, and G. Marsaja. 1996. Everyone here speaks sign language, too: A Deaf village in Bali, Indonesia. In *Multicultural aspects of sociolinguistics in deaf communities*, ed. C. Lucas, 39–57. Washington, D.C.: Gallaudet University Press.

Branson J., and D. Miller. 2002. *Damned for their difference: The cultural construction of deaf people as disabled.* Washington D.C.: Gallaudet University Press.

Bourdieu, P. 1977. *Outline of a theory in practice.* Cambridge: Cambridge University Press.

Comaroff, J. 1981. Healing and cultural transformation. *Social Science and Medicine* 15B:367–78.

Delaney, C. 1991. *The seed and the soil: Gender and cosmology in Turkish village society.* Berkeley: University of California Press.

Douglas, M. 1996. *Natural symbols: Exploration in cosmology.* London: Routledge.

Ferguson, C. A. 1996. Diglossia revisited. In *Understanding Arabic: Essays in contemporary Arabic Linguistics*, ed. A. Elgibali, 49–67. Cairo: American University in Cairo Press.

Goffman, E. 1963. *Stigma: Notes on the management of spoiled identity.* Englewood Cliffs, N.J.: Prentice Hall.

Good, B. 1994. *Medicine, rationality and experience: An anthropological perspective.* Cambridge: Cambridge University Press.

Groce, N. 1985. *Everyone here spoke sign language: Hereditary deafness on Martha's Vineyard.* Cambridge, Mass.: Harvard University Press.

Hymes, D. 1974. *Foundations in sociolinguistics: An ethnographic approach.* Philadelphia: University of Pennsylvania Press.

Jansen, W. 2000. Sleeping in the womb: Protracted pregnancies in the Maghreb. *The Muslim World* 90:218–37.

Jepson, J. 1991. Two sign languages in a single village in India. *Sign Language Studies* 70:47–59.

Johnson, R. E. 1991. Sign language, culture and community in a traditional Yucatan Maya Village. *Sign Language Studies* 73:461–74.

Khuri, F. 1970. Parallel cousin marriage reconsidered: A Middle Eastern practice that nullifies the effect of marriage on the intensity of family relationships. *Man* 5:597–618.

Kisch, S. 2000. "Deaf discourse": The social construction of deafness in a Bedouin community. M. A. thesis, Tel Aviv University [in Hebrew].

———. n.d. "Deaf discourse": The social construction of deafness in a Bedouin community. Manuscript.

Kressel, G. M. 1992. Descent through males: An anthropological investigation into the patterns underlying social hierarchy, kinship, and marriage among former Bedouin in the Ramla-Lod area. Wiesbaden, Germany: Harrassowitz Verlag.

Lane, H., R. C. Pillard, and M. French. 2000. Origins of the American deaf world: Assimilating and differentiating societies and their relation to genetic patterning. In *The signs of language revisited*, ed. K. Emmorey and H. Lane. Mahwah, N.J.: Erlbaum.

Latour, B. 1993. *We have never been modern.* Cambridge, Mass.: Harvard University Press.

Leach, E. 1969. *Genesis as myth and other essays.* London: Jonathan Cape.

Marx, E. 1967. *Bedouins of the Negev.* Manchester, U.K.: Manchester University Press.

Nelkin, D., and S. Lindee. 1995. *The DNA mystic: The gene as cultural icon.* New York: W.W. Freeman.

Ortner, S. 1974. Is female to male as nature is to culture. In *Woman, culture and society,* ed. M. Rosaldo and L. Lampher, 67–87. Stanford: Stanford University Press.

Padden, C., and T. Humphries. 1988. *Deaf in America: Voices from a culture.* Cambridge, Mass.: Harvard University Press.

Parsons, E., and P. Atkinson. 1992. Lay construction of genetic risk. *Sociology of Health and Illness* 14 (4): 437–55.

Rapp, R. 1988. Chromosomes and communication: The discourse of genetic counseling. *Medical Anthropology Quarterly* 2:143–57.

———. 1999. *Testing women, testing the fetus: The social impact of amniocentesis in America.* London: Routledge.

Scott, D. A., M. L. Kraft, R. Carmi, A. Ramesh, K. Elbedour, Y. Yairi, C. R. Srikumari Srisailapathy, S. S. Rosengren, A. F. Markham, R. F. Mueller, N. J. Lench, G. Van Camp, R. J. H. Smith, and V. C. Sheffield. 1998. Identification of mutations in the connexin 26 gene that cause autosomal recessive nonsyndromic hearing loss. *Human Mutation* 11 (5): 387–94.

Scott, D. A., R. Carmi, K. Elbedour, G. M. Duyk, E. M. Stone, and V. C. Sheffield. 1995. Nonsyndromic autosomal recessive deafness is linked to the DFNB1 locus in a large inbred Bedouin family from Israel. *American Journal of Human Genetics* 57:965–68.

Volkov, S. 1998. The deaf as a minority group and the early controversy over sign language. *Historia* 1:55–94 [in Hebrew].

NOT THIS PIG: DIGNITY, IMAGINATION, AND INFORMED CONSENT

Mark Willis

In the heyday of eugenics in the 1920s and 30s, you could not avoid a figure of speech that I will call the "litany of defectives." You would find it in college biology texts and popular magazine stories about having healthy babies. It was considered by some to be cutting-edge science, and you could run smack into it at the state fair, where stuffed guinea pigs, white ones and black ones, would be arranged on a board to illustrate Mendel's laws of inheritance. You can imagine which colors represented "pure" and "abnormal" parents and offspring. The display would be accompanied by a version of the litany that went something like this: idiocy, feeblemindedness, insanity, blindness, deafness, epilepsy, criminality, prostitution, alcoholism, and pauperism are just a few of the undesirable human traits inherited in the same way as color in guinea pigs.[1]

A more restrained iteration of the litany was recorded in *Buck v. Bell* (1927), the landmark Supreme Court decision that upheld the police power of the states to compel sterilization of mentally incompetent people housed in state institutions. Justice Oliver Wendell Holmes, the Great Dissenter, spoke for the majority when he affirmed that "heredity plays an important part in the transmission of insanity, imbecility,

&c."[2] Eugenicists in Nazi Germany, who took American legislation as a model when they enacted their own sterilization law in 1933, reduced the litany to one simple, all-encompassing principle: *lebenunwertes Leben*, or "life not worth living."[3]

These figures of speech engage what bioethicist Bruce Jennings called the "genetic imaginary," the representation of some hypothetical future life based on a selective focus on genetic information.[4] Prospective parents construct a genetic imaginary of a future child when they make choices about abortion based on prenatal genetic screening. Anyone who tries to understand the meaning of being "at risk" for a genetic disease does so by constructing a genetic imaginary of life with that disease. Even the trendy consumers who are getting their DNA scanned at commercial gene boutiques are engaging in this imaginary process.

I have spent most of my life trying to understand my own relationship with the genetic imaginary. I've spent a long time imagining what Holmes's "&c." might mean. I admit, I have an overactive historical imagination. In the 1920s, I could have been labeled "hereditary defective." Today we use kinder and gentler euphemisms. I am the carrier—some might say the victim—of two genetic diseases. A third "affliction" may be waiting in my genes. If there is an emerging genetic underclass, as Dorothy Nelkin predicted, I could run for class president or class clown.[5]

I first encountered the genetic imaginary in a darkened eye examination room when I was eighteen years old. After several years of inconclusive visits to ophthalmologists, I was referred to a retina specialist at the university hospital. Never a good sign. My father took off from work to accompany me through the long day of testing and waiting.

The retina specialist placed a row of 35-mm slides on a viewing screen in front of me. My eyes were still dazed and dilated from the rapid sequence photography that had produced the slides. I could not have seen much then, but I know now that the images resembled telescopic photographs of the planet Mars, reddish discs replete with canals (the retinal blood vessels) and a polar cap (where the optic nerve enters the back of the eye). The specialist pointed to a yellow cloud at the disk's center. "See the bull's-eye?" he asked. I did not. But I would know its shape anywhere now—it is the swirl of flashing lights at the center of my vision, the amorphous cloud through which I see the world.

The doctor told me that I had Stargardt disease, a rare form of macular degeneration that had been described in the medical literature only a few years earlier. Unlike age-related macular degeneration, the most common cause of blindness in older people, Stargardt usually began in adolescence or early adulthood. Its progression was unpredictable, and at that time, there was no way to treat it.

"You are legally blind now," the doctor said, "so you probably qualify for some kind of social services.

"It's hereditary," the doctor added, "a recessive trait that can skip one generation and show up unexpectedly in the next." He looked to my father and asked, "Is there any other history of eye disease in the family?"

My father paused and took a deep breath. Something troubling in his experience as a parent, something he could not solve for his children, began to come clear in his mind. "Mark's older sister has had eye problems ever since she was a little girl. We took her to one doctor after another, and no one could tell us what was wrong." That is how my sister Diana, then age thirty, came to be diagnosed with the same genetic disease.

Before the specialist left the room, he said, "There may not be a cure now, but there is always the hope that research will find one someday. If there is a research study, would you be willing to be in it?"

"Maybe," was all I said then.

"We don't see this very often," he added. "Do you mind if the residents take a look?" I wanted to withdraw somewhere deep inside myself, to figure out what all this meant, but I agreed. That was my first experience with informed consent.

Seven or eight young doctors in training lined up to shine ophthalmoscopes into my eyes. I began to sink after the third resident took a long, probing look. I felt like I might pass out. My father recognized my distress and stepped between the next doctor and me. "That's enough," he said. "You'll have to learn about it some other way."

My father became the guardian of my dignity then. Years later, our roles would reverse, and his quiet, decisive way of stepping in would be a powerful model for me.

Several days after my diagnosis, when I thought more about the prospect of becoming a guinea pig in a research project, an answer came clear in my mind: "Not this pig." This was my first oral formula, a mnemonic device like Homer's "wine-dark sea," a cluster of words that can summon from memory a complex array of feelings and experiences. It became my way of claiming and reclaiming the terms of an evolving relationship with the genetic imaginary.

Not This Pig was the title of a book of poems by Philip Levine and the last line of a defiant poem called "Animals Are Passing from Our Lives."[6] The poet speaks in the voice of a pig being driven to market. It can smell the butcher's block and blade. It can imagine "the pudgy white fingers/that shake out the intestines/like a hankie," but it won't fall down or squeal. Levine said the poem celebrated "digging in your heels."[7] That pig held on to its own stubborn sense of dignity, and so would I.

The next encounter with the genetic imaginary occurred twenty years later when I was thirty-nine. I was having a heart attack. As my heartbeat slipped away in the emergency room, I lost consciousness. I felt as if I'd fallen to the bottom of a well. I didn't see the apocryphal light that others have reported at the edge of death, but I heard a voice. It was my voice. It was saying, "Mark, what the hell are you doing down here?"

After a jolt of atropine, my heart started again and my eyes opened. I heard the emergency physician say, "Mark, come back."

Then a nurse ran my gurney down the hall to the cardiac catheter lab. Another nurse jogged alongside me. She held up a clipboard and explained, "This is a consent form."

"I can't see that," I said.

"That's OK." There was something gentle in her voice that I needed to hear then. "It says that you understand and agree to have a heart catheter procedure. Depending on what happens with that, you may go right into surgery. This form also gives your consent for open heart surgery if it's needed."

I don't know how many seconds it took for that to sink in. As I scrawled my signature, illegible in the best of circumstances, I knew that this might be the last decision I would ever make. In that moment, all my dignity as a human being was focused in the simple act of *making* that decision.

Now, and from this distance, you might say that the act of signing a consent form then was purely a formality, something the hospital's lawyers required in the name of risk management. Had I lost consciousness again, they would do whatever they had to do anyway, whether I signed or not. But I saw it differently. I was being given a choice, and I was determined to choose for myself.

I flirted with cardiac arrest in the catheter lab. I heard a remote voice on a speaker shout, "De-fib!" The imaging cameras swung out of the way. A nurse came toward me with defibrillator paddles. In an instant I thought, "Settle down, boy, you're about to get electrocuted."

My heart settled down. Another voice said, "Wait." The nurse was close enough for me to see her face and make eye contact, an intimacy I experience but rarely with strangers. I watched her emotions race across her face: fear, mercy, resolve to act, relief at not having to do so.

After another hour of clot-busting drugs and angioplasty, my coronary arteries were open again. I was transferred to cardiac intensive care. I felt a fleeting sense of euphoria when I found my children waiting for me there.

Eventually I realized that I had been in this hospital room before. My father had been placed in the same room after suffering ventricular tachycardia and irreversible brain damage following a heart attack. It was there, listening to the rhythmic pulse of the respirator that sustained his breathing that I began to imagine what "persistent vegetative state" means. He lived in a coma for eleven months. I became the guardian of his dignity then. Once I had to ask a hospital social worker to step out of this room when she tried to discuss end-of-life decisions as if he were not there.

Our family history of heart disease was not new to me. Several years after my father's death, I was diagnosed with a familial type of lipid disorder, which a premature heart attack now confirmed. This was my second genetic disease. I didn't just give a family history to every doctor who asked—I was re-living it. Later that night, as my heart lurched through its own arrhythmias, I heard one nurse say to another, "Keep an eye on this one."

The informed consent dialogue continued at intervals over the next two days, as my cardiologist discussed the options. He believed the angioplasty was only a temporary fix and recommended coronary bypass surgery. I agreed but needed time for my sister Diana to travel here. She was the next of kin who would serve as my guardian and proxy decision-maker.

I was talking with Diana when I lost consciousness again. When I opened my eyes sometime later, I heard her wailing in the hall. The cardiologist and a team of nurses were working on me. "Your arteries are closing up," he explained. "We're getting you ready for surgery now."

Diana was present when the heart surgeon came in to talk with me. By that time, I was as scared as I've ever been. He offered me a mild sedative. This man projected a calm, reassuring bedside manner as he explained what was about to happen.

Then he added, "There is a slight chance, maybe 1 chance in 100, that you will throw a blood clot during the surgery. This means you could have a stroke. Worst case scenario . . . massive brain damage. We could do all the heroics with life support, but maybe the best thing would be to just let you go."

The sedative was beginning to take effect, and I struggled to stay awake. Was he really saying that? I understood that I might end up trapped in a life not worth living. *Lebenunwertes Leben*? Many people say they would make that rational choice rather than prolong life in a persistent vegetative state. That's supposed to be the choice with dignity. But at that moment, vegetable or not, I wouldn't say yes to death.

"I want a chance," I said. "I want to live." Then I fell asleep, trusting that my sister understood what I meant.

Several years ago, the phone rang at home after dinner one night. My son Brendan answered it, ever wary of the telemarketers who swarm like mosquitoes as the sun goes down and people begin to relax. Brendan was fourteen then, and he had my permission to goof on telemarketers however he wanted. Something in the voice on the other end made him restrain himself.

When I picked up the phone, Dr. X said, "Mr. Willis, I'm waiting for you to return the signed consent form." Dr. X was a young physician-scientist at a prestigious research university who was searching for the gene that caused my eye disease. We had talked on the phone once before, when I was in my office at another university. He wanted me and my family to join his research study. I was the point of entry for an entire "pedigree," as geneticists say.

"I haven't had a chance to read it yet," I said a little sheepishly. My life is filled with stacks of unread papers, preserved through time and neglect like the geologic record. I trust the stratification, and maybe someone will come along who will read some of it to me. I felt that I could be honest with Dr. X. After all, he was devoting his career to curing my disease.

"It's a long, technical document," I continued, a touch of defensiveness in my voice. "The type is smaller than what I can manage these days."

"You could just sign it. All we want is a blood sample."

"No," I said. "I need to read it. The part about 'informed' is just as important as the part about 'consent.'"

"When will you be able to read it? We don't have a lot of time." He sounded impatient and a little angry now. He repeated his recruitment pitch. "We're close to identifying the gene for your problem. You know how important this research is. In a year we'll have a screening test. Then you can find out if your son will get Stargardt's. The screening will be free, of course, for participants in the study."

I was getting angry now, too. I wasn't ready to bring Brendan into this or debate the social risks of genetic screening. "Look," I said shortly, "I need to read your document before I consent. I need to read *all* of it. Can you send me a tape or read it to me over the phone?"

"Can't your wife read it to you?"

"My wife?" I could hear stigma piling on top of stigma like a litany: blind, genetic defect, single parent. . . . Was this an audition for the Jukes family? Getting a grip, I said, "I don't have a wife. Even if I did, that's not the answer. If you're going to do research with blind people, you have an obligation to provide them with reasonable accommodations."

"We don't have money for that," Dr. X said.

"Well, then, not this pig."

I wanted to say that, but something in this busy scientist's voice made me hold back. He didn't even know that we were negotiating.

"Good night, doctor," I said. "I'll read it when I can." I never did, and my family has not yet donated its genotype to science.

In the name of full disclosure, I should make it clear that I am not a twenty-first-century Luddite who categorically opposes the advance of science. After all, I make my living as a science writer at a medical school. I am a disability rights activist, but I cannot muster an "undifferentiated moral condemnation" of the medical model, as it is represented sometimes in the discourse of disability studies.[8] I was interested in joining Dr. X's study out of intellectual curiosity. I've volunteered for several human research projects, and none of them posed a risk to my well-being or dignity. I have imagined the circumstances in which I'd take experimental risks in the management of my heart disease, and I know I could make the decision without much time to think about it.

Hoping for an experimental cure for my eye disease, however, is not even a blip on my sonar screen. A geneticist who has heard my stories asked me once about this difference in attitudes. The simplest answer is this: Unlike heart disease in both its acute and chronic dimensions, I do not experience vision loss as a disease. It is a different way of perceiving the world, and it is rich with its own sensory skills and sweet satisfactions. I think of myself as socially blind; the deficits associated with my blindness result more from society's limitations than from a disease process active in my body.

Instead of looking to DNA for answers, I've lived my life by improvising on the genome. This is the story that reaches beyond the script encoded in my genes. It's the story I make up as I go. It's a body of knowledge and experience with making adaptations and negotiating accommodations. I am not likely to enter someone else's construction of the genetic imaginary unless it makes room for this body of lived experience.

In my case, it is a family story, too. My sister has been a guide and mentor throughout my life. Today she is a master teacher with a special education classroom devoted to early childhood intervention, and she brings to it a disability activist's commitment and a grandmother's wisdom. Together, we have more than 100 years of experience with our disability, and we put our faith in that.

About a month after Dr. X called, Brendan was reading to me from an essay in Mikhail Bakhtin's *The Dialogic Imagination*. The abstruse prose studded with Russian linguistic terms was rough sledding, but he agreed to try it after the audiotape broke. By the age of ten, Brendan could read anything, and he often read aloud to me. I thought in passing that he could have read Dr. X's consent form to me, until I remembered that it involved his genes as much as mine.

Something in Bakhtin's radical social ideas about language resonated with Brendan. He stopped reading and confided to me that when he grew up, he wanted to be a rebel rather than middle class.

"Me too," I said. "On the surface, I have all the trappings of a middle-class life—credit cards, a house, a college education. But I don't think anyone's vision of the middle class includes living with a disability."

Then my son revealed a glimpse of his own construction of the genetic imaginary. "You know, dad, I used to think it was sad that you were going blind. Now I think it is just the way you are. I can't imagine you any other way."

When I was a child, and there was no limit to my reading, a biography of Justice Oliver Wendell Holmes captured my imagination. *Yankee from Olympus* cast the Great Dissenter in a heroic mold, and it led me to his eloquent dissenting opinions on freedom of thought and expression.[9] By the 1960s, these minority opinions had prevailed as precedents of First Amendment law. Justice Holmes argued in the 1919 case of *Abrams v. United States* that government must not ban or punish seditious speech for fear of its consequences. "The ultimate good desired is better reached by free trade in ideas," he wrote; "the best test of truth is the power of thought to get itself accepted in the competition of the market."[10] I put this in my own words and took it as a credo: Speech, no matter how extreme, must be trusted to the free marketplace of ideas.

Yankee from Olympus did not mention *Buck v. Bell*. I could never have imagined as a child that one day I would hear the decision's most notorious argument—"Three generations of imbeciles are enough"— and feel betrayed.[11]

Justice Holmes never met Carrie Buck, the plaintiff in this case. He accepted her diagnosis as an imbecile and consigned her to the litany of defectives on the authority of written testimony from experts.

He had to imagine what a life of imbecility might mean, and he imagined life not worth living. "It is better for all the world," he wrote, "if instead of waiting to execute degenerate offspring for crime, or to let them starve for their imbecility, society can prevent those who are manifestly unfit from continuing their kind."[12]

That was Justice Holmes's construction of the genetic imaginary. The U.S. Supreme Court codified it as the law of the land and stamped it with the highest authority to enforce decisions that led to the sterilization of thousands of Americans.

The judicial weight of *Buck v. Bell* underscores the first point I want to make. At the intersection of law, medicine, and science, institutions wield great power to shape both the information and the decisions we make in the informed consent process. According to Bruce Jennings, "We must not underestimate the power of science and technology to colonize and dominate the contemporary imagination."[13] In other words, when we make decisions based on informed consent, especially in circumstances when our autonomy is most vulnerable, the marketplace of ideas may not be as free as it should be.

My final point emphasizes the experience of families in this process. When Dr. X asked me to join his genetic study, I had legal responsibility for me, for my son who was a minor, and for my mother. She had Alzheimer's disease (that is the third affliction that may be waiting in my genes), and I was her guardian and medical decision-maker while she lived in a nursing home. Like Justice Holmes, I could have rendered a pragmatic, paternalistic decision for three generations. I was prepared to give my DNA sample to Dr. X, but I knew I would wait for my son to make his own decision about it, no matter how long that might take. I needed to read the informed consent document carefully to know how it would protect his privacy and mine.

By the time Dr. X called me at home, I also knew that I could not make this decision for my mother. It was just a blood sample, as the doctor said, and she might have forgotten all about it five minutes after the needle stick. Or her memory might have stuttered about it in ways I could not know. When she was competent to make her own decisions, she would have signed a consent form had I asked her, but she never would have done it without my influence.

I could say I was the guardian of her dignity then, but really, family roles reversed one more time. At the end of her life, my mother was demented and paralyzed and nourished with a gastric feeding tube. Her final stroke took her swallow function and ability to speak. Some people would say that is a life not worth living. In her presence, though, I was reminded of Emerson's simplest statement of faith: "I believe in the still small voice."[14] Her serenity and acceptance of her body gave me one more lesson in the awesome dignity of living life out however it unfolds. With one moveable hand, my mother remembered how to take a flower and raise it to her nose to sniff. This is how we conversed.

We were improvising on the genome. Life's improvisation remains an unfinished project for all of us. I believe each of us can play a part in it no matter how small the voice. For that, three generations are not enough.

NOTES

1. See Daniel J. Kevles, *In The Name of Eugenics: Genetics and the Uses of Human Heredity* (New York: Knopf, 1985), ch. 4.

2. 274 U.S. 206 (1927).

3. See Robert Jay Lifton, *The Nazi Doctors: Medical Killing and the Psychology of Genocide* (New York: Basic Books, 1986), 20. Lifton translates *lebenunwertes Leben* as "life unworthy of life."

4. Bruce Jennings, "Technology and the Genetic Imaginary: Prenatal Testing and the Construction of Disability" in *Prenatal Testing and Disability Rights* ed. Erik Parens and Adrienne Asch (Washington, D.C.: Georgetown University Press, 2000).

5. Dorothy Nelkin, "The Social Power of Genetic Information" in *The Code of Codes: Scientific and Social Issues in the Human Genome Project* ed. Daniel J. Kevles and Leroy Hood (Cambridge, Mass.: Harvard University Press, 1992), 190.

6. Philip Levine, *Not This Pig* (Middletown, Conn.: Wesleyan University Press, 1968).

7. Quoted in David Remnick, "An Interview with Philip Levine," *Michigan Quarterly Review* 19 (1980): 382–98.

8. See Lifton, *The Nazi Doctors,* 503, on his struggle with bearing witness without losing critical focus through "undifferentiated moral condemnation" of medical killing in the Holocaust.

9. Catherine Drinker Bowen, *Yankee from Olympus: Justice Holmes and His Family* (Boston: Little Brown, 1944).

10. 250 U.S. 630 (1919).

11. 274 U.S. 207 (1927).

12. Ibid.

13. Jennings, "Technology and the Genetic Imaginary," 141.

14. Quoted in Robert D. Richardson, Jr., *Emerson: The Mind on Fire* (Berkeley: University of California Press, 1995), 158. The Biblical source for the phrase is the story of Elijah and the still small voice (1 Kings 19). Richardson places Emerson's statement in the context of his affinity with Quakerism in the 1830s.

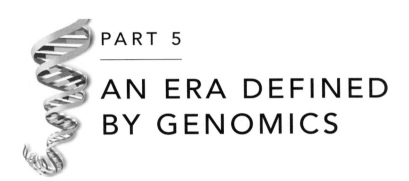

PART 5

AN ERA DEFINED BY GENOMICS

INTRODUCTION

The final two essays in this collection look at the broad social and political consequences of genetic research and the place of disability in the popular imagination and in government policy. Both challenge the optimism that characterizes the public and scholarly statements of geneticists. They argue that American society's attitudes toward and treatment of people with disabilities are not reassuring in regards to how genetic knowledge and technical advances will be used in the future, or are being used today, as genetic screening, in vitro fertilization, and selective abortion become common tools to ensure that children will have "desirable" genetic traits.

Deaf scholar Christopher Krentz uses analyses of Mary Shelley's novel *Frankenstein*, published in 1818, and a 1997 film, *Gattaca*, to discuss modern society's desire to achieve perfection. He argues that both these artifacts of popular culture address important issues, particularly the "line between desirable [scientific] progress and folly," which transcends short historical periods and that must be constantly reexamined. Though critical of parts of both *Frankenstein* and *Gattaca*, Krentz concludes that they demonstrate the importance of "embracing difference and disability."

Bérubé expands on Krentz in a number of ways and pushes his critique further. He directly confronts the thorny question of which differences and which disabilities should be embraced, speculating, for example, that a freely democratic discussion might produce a popular majority—though not a consensus—that it was morally defensible to abort a fetus with Down syndrome but not one with deafness. He insists, as well, that questions like this must be addressed in a public and, importantly, a non-theological way. They cannot be left to scientists, for "contemporary professions of good faith among geneticists are no guarantee that they've learned the lessons of history." Bérubé insists on the importance of politics. "Disability is always and everywhere a public issue, a matter for public policy," he writes, concluding that "the experience of disability" must be central to politics in "the age of genomics."

FRANKENSTEIN, GATTACA, AND THE QUEST FOR PERFECTION

Christopher Krentz

Fifty years ago, Francis Crick and James Watson discovered the structure of DNA, helping to launch a scientific revolution that is affecting many aspects of our lives. The list of developments is both exciting and a bit unsettling. Perhaps most encouraging, scientists agree that genes can yield cures to such diseases as Parkinson's, Alzheimer's, cancer, and malaria. Already, researchers are using genetically manipulated bacteria to make human insulin, growth hormone, and a vaccine for hepatitis B. At the same time, genetics has given rise to medical procedures like in vitro fertilization and prenatal screening, which some people worry will lead to a new eugenics that seeks to improve the human race through selective breeding. Genetics has produced a boom in the production of life-enhancing drugs, extending the quality and length of human life and bolstering the economy. Yet the discovery of DNA has also made it easier for some groups to try to create new bacteria and viruses for biological weapons. Genetics has changed agriculture, making it possible to raise more crops more efficiently; currently two-thirds of the soybeans and one-fourth of the corn grown in the United States is transgenic.[1] However, some critics contend that these new foods have not been adequately tested; especially in Europe and Asia, resistance to such foods is on the rise. Scientists have cloned animals, including Dolly the sheep, CC the cat, and, more recently, an endangered species called the Javan banteng. They have also cloned plants and microorganisms, even as legislators and ethicists debate the desirability of such actions. One wonders when humans will be cloned, if they haven't been already. Finally, in November of 2002, the *Washington Post* reported that a group of scientists was working on creating a single-celled organism that is new form of life, which adds to the wonder and uncertainty of the age.[2]

Where does all this leave us? If modernity means living through constant change, as the cultural historian Marshall Berman has suggested, then genetic engineering places us firmly in modern times. Berman states that to be modern is "to find ourselves in an environment that promises us adventure, power, joy, growth, transformation of ourselves and the world—and, at the same time, that threatens to destroy everything we have, everything we know, everything we are."[3] His words aptly sum up the promise and threat of tinkering with the genome.

One way to approach these sometimes overwhelming transformations is to look at how they are portrayed in literature and film. Our stories can be powerful, expressing our fears and desires, depicting the details of everyday experience as well as the deeper meanings of life, and acting as tools for social change. In short, literature and film can teach us about ourselves. When I think of genetic engineering, two creative works immediately come to mind: Mary Shelley's 1818 novel *Frankenstein*, which is still one of our most dominant myths about biological experimentation, and Andrew Niccol's 1997 film *Gattaca*, which was well-received by critics but did not do all that well at the box office. Although separated by almost 200 years, *Frankenstein* and *Gattaca* contain striking similarities. While not overtly against science, both use complex figures of disability to caution against unthinking scientific innovation. Mary Shelley's Creature and the protagonists in *Gattaca* must struggle to come to terms with a world that rejects their physical makeup. In these works, disability often makes characters sympathetic, more colorful, and more human. Yet despite this positive use of physical difference, in the end, both the novel and the film move to erase disability, showing how deeply the desire for perfection can run.

When Mary Shelley wrote *Frankenstein* in her late teens (she published the novel when she was just twenty years old), she drew upon a wide range of sources, including contemporary developments in science. One such advancement was the discovery of electricity. In public exhibitions, demonstrators applied electrical stimulation to the bodies of recently executed criminals. When the bodies shook and moved, some people wondered if electricity could give life to dead matter—an idea that Shelley takes up in her book. In addition, as Jon Turney points out, she addresses the speculations of Erasmus Darwin, a prominent physician who suggested that people might one day scientifically create life. She also reproduces some of the chemistry rhetoric of her day, and Victor's foraging for parts for the Creature has parallels in the early-nineteenth-century practice of exhuming and dissecting corpses. Turney concludes that Shelley's novel presents a "surprisingly realistic composite picture of contemporary science."[4] Yet her attitude toward this science is anything but straightforward.

Even before the narrative begins, Shelley expresses ambivalence toward science and the pursuit of knowledge and power. The original 1818 title page gives the full title of the novel as *Frankenstein; or, the Modern Prometheus*. Frankenstein, of course, refers to Victor Frankenstein, the brilliant student who discovers the secret to life (the Creature remains unnamed in the novel). When Victor decides to try to make a man, he uses dead human and animal parts, mixing species rather like scientists today may mix the organs or genes of one animal with another. He succeeds in creating a new life-form unlike anything on earth. The subtitle links Victor with Prometheus, the Titan in classical mythology whom, in some accounts, is asked by Zeus to create humans. We more often remember Prometheus for stealing fire from the gods and giving it to people, for which Zeus punished him by chaining him to a rock and having a bird devour his liver every day. The subtitle thus connects Victor Frankenstein not only with the creation of life, but also with rebellion against God and undergoing torture as a result. Some critics have read Shelley's portrayal of Victor as a wry take on inflated masculine aspirations in general, and on her ambitious poet-husband, Percy Shelley, in particular.

Shelley further complicates the heroic nature of the title by including an epigraph that suggests the Creature's point of view. Even on the title page, she offers more than one perspective, a tactic she develops throughout her complex, multilayered novel. In the epigraph from *Paradise Lost*, Adam addresses God in the Garden of Eden:

Did I request thee, Maker, from my clay
To mould me man? Did I solicit thee
From darkness to promote me?

Adam's confused words correspond to the Creature's helplessness when
he finds himself alive but without any love or support. By implicitly com-
paring the Creature to Adam, Shelley suggests that the Creature has
human attributes, that in spite of his transgressive body, he in some
ways resembles the first man. Yet we cannot say exactly what the Creature
is. Again, we could compare him to the genetically engineered organisms of today. If scientists insert human genes into nonhuman organisms, how many human genes does that organism need before we
consider it human?

The Creature remains ambiguous and elusive, treading the boundaries between identity categories. It is not even entirely clear what he
looks like. Victor tells us that the Creature has "gigantic stature" and
is "about eight feet in height."[5] Shortly after the Creature comes to life,
Victor says,

His limbs were in proportion, and I had selected his features
as beautiful. Beautiful!—Great God! His yellow skin barely
covered the work of muscles and arteries beneath; his hair was
of lustrous black, and flowing; his teeth of pearly whiteness;
but these luxuriances only formed a more horrid contrast with
his watery eyes . . . shriveled complexion and straight black
lips. (56)

Despite this description of specific attributes, from yellow skin to watery
eyes, it remains difficult to imagine what the Creature looks like overall. What is clear is that the people who see the Creature find him
hideously ugly.[6] When the Creature comes to life, he stretches out his
hand to Victor like a newborn baby reaching for his parent. However,
his ghastly appearance causes Victor to flee in horror.

The Creature's appearance makes him into a figure of disability.
Scholars have interpreted the Creature in any number of ways; they
have seen him as representing women or racial minorities, as a "natural man," and so on. These readings all may have some validity, but
they usually bypass the Creature's own self-description: he views him-

self as disfigured. When he first catches a glimpse of himself in a pool, he reacts with terror, responding with "the bitterest sensations of despondence and mortification" to what he calls his "miserable deformity" (110). He subsequently laments that he is "endued with a figure hideously deformed and loathsome" (116–17) and says that "one as deformed and horrible as myself [has] defects" (139). Why does he see himself this way? The Creature's words raise the question of whether he is inherently ugly, or whether this ugliness is something socially determined. Aside from his appearance, the Creature seems to have no physical limitations at all: He is exceptionally intelligent, quickly educating himself and articulately pleading his case; he has superior strength, speed, and is able to live in icy terrain on nuts and berries. We can imagine that, in a community of similar creatures, he would be happy and well adjusted. But as the only one, surrounded by humans, he comes to judge himself by human standards of beauty and finds himself lacking.

The humans the Creature encounters almost always reinforce his self-loathing. When he awakens, Victor's appalled reaction mirrors many parents' initial response to having a child with a disability, and he leaves the Creature with no parent, love, identity, or community. Wherever he goes, the Creature's efforts to find acceptance are rejected because of his "deformity" (110). He is shot at, beaten, and condemned. The only character who treats him respectfully is the blind man De Lacey, who converses with the Creature as he would with any other person. When De Lacey's family returns, they hit the Creature in fury and drive him away. Ironically, it takes a disabled character who can't see to perceive the Creature's goodness. In this book, disability often serves as a subtle trope for humanity.

Shelley puts us, as readers, in something of the blind man's position; we do not ever *see* the Creature's horrible aspect, but we do witness his innocence, genuine emotions, and vulnerability, which initially makes him a sympathetic character. He expresses the most human hopes: at one point, he says, "My heart yearned to be known and loved by [the De Lacey family]; to see their sweet looks directed towards me with affection was the utmost limit of my ambition" (128). The Creature voices the same needs for acceptance, love, and community that we all have; he is a social animal unjustly denied any chance of socialization. By making the Creature speak for himself, Shelley helps us to see him as a victim with real human emotions. Ironically, he sometimes comes off as more human than the self-centered Victor, who flees from his creation at its birth and who subsequently refuses to make him a mate. In these ways, the Creature's disability actually helps to humanize him, to make him into a figure of oppression with which it is easy to empathize. Mark Mossman, a disabled scholar, recalls powerfully identifying with the Creature when he first read *Frankenstein* at age eighteen: "I felt all the resentment of the creature, the anger, the isolation, the loneliness. The creature was the ultimate victim of stereotyped oppression, of a disabling construction of 'ugliness,'" he writes.[7] In the early parts of his story, the Creature's disability makes him an underdog, an everyman, an imperfect protagonist competing against a bigoted system. It paradoxically makes him more human.

When the Creature in anger begins to murder Victor's family and friends to make Victor as isolated as himself, it may at first seem the fulfillment of his hideous appearance. Shelley skillfully plays off a long line of disabled villains, from Richard III on down; we almost expect the hideous figure to prove evil, for an ugliness of skin to translate into an ugliness of character. People must have been correct all along to fear him, we think, since he behaves in such monstrous ways. However, while not denying the Creature's responsibility for his actions, Shelley is careful to plant clues that Victor is equally culpable. When Victor rejects the Creature, he sets the Creature on the path to become a monster. Would the Creature have been driven to murder if Victor had taken care of him like a responsible parent (or god)? This modern Prometheus, Shelley seems to suggest, is too shortsighted, too filled with his own needs and ambitions, to treat his creation in a responsible way.

If the Creature represents a disabled body after his birth, Victor proves something of a eugenicist. When the Creature finds Victor and demands that Victor make him a mate, Victor in fear agrees. Yet, before he finishes the mate, Victor destroys her; he does not want to be responsible for a threatening, inferior race of beings. Here Shelley anticipates the eugenics movement of the late-nineteenth and early-twentieth centuries, which sought to improve the human race through selective breeding. In 1883, Alexander Graham Bell published *A Memoir upon the Formation of a Deaf Variety of the Human Race*, which argues against deaf intermarriage out of the same kind of fear that Victor has.

Although Victor and the Creature may seem diametrically opposed—creator versus created, human versus nonhuman, and eugenicist versus

THE QUEST FOR PERFECTION

disabled—in the final stages of the novel, Shelley makes them increasingly similar. They begin to echo each other's words, they are both isolated outcasts, and both are determined to gain revenge on the other. The line between them blurs, and we see even more clearly how connected they are to each other.

In the end, Victor does not learn from his experience. "I have myself been blasted in these hopes," he says of his scientific questing shortly before he dies, "yet another may succeed" (210). These are his last words. He does not realize he may have exceeded the limits of acceptable science, just as he does not truly accept his failure as a parental figure. Victor actually expresses the true moral of the story near the beginning of the novel, although chronologically toward the end. In a passage that has not received the attention it deserves, he states, "We are unfashioned creatures, but half made up, if one wiser, better, dearer, than ourselves—such a friend ought to be—do not lend his aid to perfectionate our weak and faulty natures" (27). His use of the word *creature* reminds us that we all potentially share the Creature's fate, that we all are either disabled in one way or another or can easily become that way, and need the love, acceptance, and support of others to thrive. The Creature's disability is metaphorically all of ours, Shelley seems to suggest. This humanizing mark of difference reveals the dangers of the arrogant use of science and disregard for other life-forms.

How have subsequent versions changed Mary Shelley's story? The first published image of the Creature, which appeared in the 1831 edition of the novel, portrays him as muscular, bewildered, and quite human. However, film versions have often emphasized the Creature's monstrosity over his human aspects.[8] In the films we are outside the Creature and see his ugliness just as the characters in the novel do. Furthermore, in many films, the Creature does not speak, which robs him of the impressive eloquence that Mary Shelley originally gave him and may make him seem less human. As the first screen creature in Edison's 1910 silent film, Charles Ogle presents an almost comically garish monster. The most famous portrayal remains Boris Karloff's depiction in the 1931 movie *Frankenstein*. Karloff does not talk in this film, and the Creature appears primarily as a monstrous menace. The producers changed Shelley's story in another way. They introduced the assistant Igor and gave a biological reason for the Creature's malicious behavior: He is given the brain of a criminal. Instead of Victor's rejection of the Creature, his failure as a parent and god, it is this biologically "unfit" brain that explains the Creature's murderous actions. This version coincides with the eugenic spirit of the 1930s. More recently, in his 1994 film *Mary Shelley's Frankenstein,* Kenneth Branagh attempts to recapture the spirit of Shelley's novel. Appearing as a quite human-sized Creature, Robert De Niro speaks, albeit haltingly. To me, all the film versions remain substantially different from the novel simply because of the form: we are outside the Creature, without much access to his consciousness, and Shelley's subtle message about the humanizing force of disability is frequently lost.

For all these permutations, the *Frankenstein* myth remains a central way that we respond to biological innovation. For example, today opponents of genetically engineered crops often refer to them as "Frankenfoods," implying that such foods are the dangerous result of the same irresponsible, careless science that Victor practices. Similarly, a British cartoon from the *Guardian* depicts a meek scientist holding a test tube who is surprised to find a Karloff-like Creature towering behind him. "Hello, Mum!" the Creature says, returning us to some of the questions posed by Shelley's novel: can we be responsible parents for the new life-forms we create, even if we create something that's different from what we expect?

Andrew Niccol's 1997 film *Gattaca,* which stars Ethan Hawke, Jude Law, and Uma Thurman, offers a more recent take on genetic issues. The action of the film takes place in the "not too distant future," in a society where genetics is dominant. In this world, you apply for a job by giving urine sample. A woman gives a man a strand of her hair for a DNA test and says, "Let me know if you're still interested." Pianists are genetically engineered so that they can play pieces that can only be played with twelve fingers. Fittingly, the name *Gattaca* is derived from the four letters of DNA code: GTCA.

In the society of *Gattaca,* parents can choose the characteristics of their offspring, if they have the money. Early in the film, a couple meets with a genetic counselor about their next child. The counselor reports that several embryos have been fertilized with the father's sperm in vitro. They have been screened for "critical dispositions to any of the major inheritable diseases," as well as for "any potentially prejudicial conditions, [including] premature baldness, myopia, alcoholism and addictive susceptibility, a propensity to violence, obesity, et cetera."[9] The parents have specified "hazel eyes, dark hair, and fair skin." The only thing that remains to be determined is the candidate's gender. After the parents select male, they wonder if it might not be better to leave some things to chance. The counselor smilingly dissuades them. "We want to give your child the best possible start," he says. "Believe me, we have enough imperfection built in already. . . . And keep in mind, the child is still you. Simply, the best of you." Such genetic manipulation may not be as far-fetched as it seems. In Australia, fertility doctors reportedly used genetic screening to dispense with several embryos that had a gene for deafness, retaining the ones that did not.[10] Moreover, in some cases, fertility clinics in the United States now have the go-ahead to let couples choose the gender of the baby by selecting one embryo and disposing of the rest.

In *Gattaca*, genetic engineering leads to a new kind of prejudice, as people are divided into two groups. "Valids" are members of the elite: they have been born through genetic engineering and have the power and privilege in this society. "Invalids" are those who are born naturally, the traditional way, without genetic engineering. They are consigned to the underclass. While the word is pronounced "In-valid" in the film, it closely resembles "invalid," with its connotations of disease and disability. People born without genetic engineering are in a sense disabled. Just as the Creature is discriminated against because of his appearance, the Invalids in *Gattaca* endure prejudice because of their DNA readings. (It's worth noting that the genetic counselor is African American; in this society, it seems, racism and other traditional -isms have been supplanted by "gene-ism," discrimination based on genetic code.) The fate of Invalids is determined at birth, by their genetic makeup, and they can do little to overcome it.

Like Shelley's Creature, the protagonist of *Gattaca*, Vincent (Ethan Hawke), is a disabled figure fighting the odds. He is an Invalid, and his disability makes it easy for us to identify with him. As soon as he's born, the nurses analyze his DNA and predict his future capacities; in his case, this includes the fact that he has a ninety-nine percent chance of dying of a heart disorder when he is 30.2 years old. Because of his genetic makeup, he can only work as a janitor, but he dreams of flying into space, something only Valids can do. No matter how hard he prepares, his genetic code dooms him to defeat. So he decides to try to beat the system. He goes to a genetic broker, who introduces him to Eugene, a Valid who has fallen on hard times and has a valuable genetic identity for sale. The broker explains, "His credentials are impeccable. . . . The guy's practically going to live forever. He's got an IQ off the register. Better than 20/20 in both eyes. And the heart of an ox. He could run through a wall . . . if he could still run." At that point, Eugene (Jude Law) appears in a wheelchair and lights a cigarette.

Despite his perfect double-helix, Eugene has a disability too. He apparently did not win a swimming race, placing second instead of first, and the disappointment led him to step in front of a car. Eugene shows that there is a gap between the genetic reading and actual life. In this respect, the film proves prescient. The Human Genome Project revealed that humans have a mere 30,000 genes, or just 5,000 more than the mustard cress plant in terms of genetic complexity. As Lennard J. Davis points out, "No one gene determines the course of a human life . . . genes alone will not tell the story."[11] For Eugene (whose name literally means "good gene"), being less than perfect proves intolerable. By the time we meet him, he's alcoholic, in a wheelchair, and has largely given up on life. He's also a colorful, entertaining character. The disabled characters like Vincent and Eugene seem much more alive than their able-bodied counterparts, whose supposed perfection usually makes them appear boring and sterile. Director Niccol makes us root for the disabled characters, who expose the injustice of the system and are the most real.

Gattaca's resolution shows that genetics does not explain all of what makes us human. Vincent winds up paying Eugene for his genetic identity, which he assumes to elude discrimination. He dyes his hair, wears colored contact lenses, and "borrows" Eugene's blood, urine, hair, and skin cells. Now he can pass the urine tests and enter the competitive space training program. After successfully completing the urinalysis, he returns jubilantly home to tell Eugene the good news. "I made it," he says, from his place on the spiral staircase that dominates the scene. Eugene, seated spiritlessly in his wheelchair at the bottom of the stairs, only replies, "Of course you made it." The spiral staircase has the shape of the DNA double-helix (just as, in the scene with the genetic counselor, the boy Vincent is playing with a DNA model on the floor. The film is filled with genetic visual imagery). In this scene, Vincent is "climbing" his DNA; his ambition helps him to beat the system, pass the tests, and achieve his dream of going off into space. The film's tag line reads, "There is no gene for the human spirit." Vincent's spirit refuses to bow to his weak genetic code; like the Creature, he strives to defy the odds against him, and unlike the Creature, he succeeds. Eugene, on the other hand, has given in to despair; when Vincent rockets into the stratosphere at the end of the film, Eugene commits suicide in a furnace.

Gattaca, like *Frankenstein*, thus uses disability to make its protagonists more human and real and expose the flaws in a system that rewards homogeneity and perfection. Yet the way both works finally deal with disability is complicated. In the end, they both purge disability, concluding with what Sharon L. Snyder has called a "cure-or-kill mindset."[12] By having their disabled characters commit suicide or go to great lengths to pass successfully as normal, Shelley and Niccol effectively erase disability, revealing how deep and persistent is our culture's desire to achieve perfection. After Victor dies in *Frankenstein*, the Creature vows to go off and burn himself on a funeral pyre. We never actually see him perish, which adds to the power of the story; yet his promise conforms to many other narratives where the threatening disabled presence is ultimately removed. Similarly, in *Gattaca*, the man in the wheelchair, Eugene, dies in flames, in effect cleansing the world of another transgressive body. Such eliminations have disturbing connections to eugenics, to a desire to cleanse the world of disability and difference. Although Vincent survives and achieves his dream, he does so by passing as able, as Valid; only by making his disability invisible does he succeed.

These works seem ambiguous, to have it both ways. They show the humanizing effects of disability and the decidedly dehumanizing effects of societal prejudice and discrimination. Yet it is easier to get rid of a problem body than to change a system, and both *Frankenstein* and *Gattaca* provide conclusions that the majority of readers and viewers (if not disability advocates) will find satisfying. Perhaps this ambivalent depiction is accurate. Most of us are probably prepared to recognize disabled people as full human beings and to decry social bigotry. But how many of us, if we could, wouldn't choose to give our children better than 20/20 eyesight? If you had to choose between an embryo with a gene for dwarfism and one without, which would you pick? If pregnant women learn through prenatal screening that their fetus has a disability, it's not hard to imagine that at least some of them would have an abortion. We as a culture remain uncomfortable with disability, even if disability is part of being human. *Frankenstein* and *Gattaca* warn of the dangers of this yearning for perfection even while they reinforce it in the end.

It is easy for many people to dismiss these works as alarmist, against scientific inquiry, or quaint and overly dramatic. However, they raise important questions that we all need to consider regularly to proceed in responsible fashion. Will our quest for perfection go too far? Can we control nature? Where is the line between desirable progress and folly? Both the novel and film also celebrate acceptance and connection with others. A large part of the tragedy of the Creature and Eugene is not that they are disabled, but that they are alone. Vincent, on the other hand, achieves his dream because of help from others. At the end of the film, a doctor catches Vincent as an Invalid but lets him board the spaceship anyway because the doctor has an ambitious Invalid son. For all his efforts, Vincent can't do it alone. He needs the assistance of his friends, Eugene, his love interest Irene, and the doctor, which takes us back to Victor's comment: We are all "unfashioned creatures," and like Vincent, we need the aid of friends to perfect "our weak and faulty natures" (27). *Frankenstein* and *Gattaca* show the importance of taking pride in who we are, of embracing difference and disability, and adjusting our attitudes to ensure that everyone is accepted, regardless of their appearance or genetic code, and that everyone is given the chance to succeed. After all, disability is part of being human.

NOTES

1. Barry Commoner, "Unraveling the DNA Myth: The Spurious Foundation of Genetic Engineering," *Harper's Magazine* (February 2002): 45.

2. Justin Gillis, "Scientists Planning To Make New Form of Life," *Washington Post* (21 November, 2002): A1+.

3. Marshall Berman, *All That Is Solid Melts into Air: The Experience of Modernity* (New York: Simon and Schuster, 1982), 15.

4. Jon Turney, *Frankenstein's Footsteps: Science, Genetics, and Popular Culture* (New Haven and London: Yale University Press, 1998), 23. Turney offers an extremely helpful exploration of how *Frankenstein* and other fic-

tional stories relate to scientific developments over the past 200 years. However, he bypasses the disability angle.

5. Mary Shelley, *Frankenstein: Or the Modern Prometheus* (London: Lackington, Hughes, Harding, Mavor, and Jones, 1818; New York: Penguin, 1992), 52.

6. For more on the Creature's ugliness, see Denise Gigante, "Facing the Ugly: The Case of *Frankenstein*," *ELH* 67 (2): 565–87.

7. Mark Mossman, "Acts of Becoming: Autobiography, *Frankenstein*, and the Postmodern Body," *Postmodern Culture* 11 (3): 5–6.

8. For a variety of images of the Creature, including screen versions, see David J. Skal, *Screams of Reason: Mad Science and Modern Culture* (New York: Norton, 1998), and Turney, *Frankenstein's Footsteps*.

9. *Gattaca*, DVD, directed by Andrew Niccol (1997; Culver City, Calif.: Columbia Tristar Home Entertainment, 2003).

10. Rick Weiss, "Screening Embryos for Deafness," *Washington Post* (14 July, 2003): A6. Somewhat ironically, Andrew Niccol, the director of *Gattaca*, is Australian himself.

11. Lennard J. Davis, *Bending over Backwards: Disability, Dismodernism, and Other Difficult Positions* (New York and London: New York University Press, 2002), 17–18.

12. Sharon L. Snyder, "Infinities of Forms: Disability Figures in Artistic Traditions," *Disability Studies: Enabling the Humanities*, ed. Sharon L. Snyder, Brenda Jo Brueggemann, and Rosemarie Garland-Thomson (New York: Modern Language Association, 2002), 180–81.

DISABILITY, DEMOCRACY, AND THE NEW GENETICS

Michael Bérubé

In this essay, I will try to suggest ways of thinking about biotechnology and disability that are compatible with democratic values. I believe there is no disputing that we have entered an era defined by genomics, an era in which our capacity for manipulating genetic material will determine what it means to be human—and in which, as I will argue, our deliberations about what it means to be human must guide our capacity for manipulating genetic material. In saying this, of course, I am presuming that political deliberations can in fact determine the scope and direction of scientific research; I am also presuming that democratic values should prevail with regard both to the procedures and the substance of such deliberations. Affirming at the outset one's commitment to deliberative democracy in such matters does not determine their outcome: it remains to be argued, for instance, whether or not it is consonant with democracy to permit prospective parents to engage in genetic selection for Huntington's disease or for myopia, or to prohibit potential genetic therapies for Alzheimer's disease or multiple sclerosis. Deliberative democrats can certainly come to different conclusions on these questions, based on opposing yet altogether reasonable assessments of how to understand and adjudicate competing claims with regard to individual liberties (positive and negative), ideas of distributive justice, legitimate scientific objectives, moral injunctions to prevent suffering, and beliefs about human dignity and nascent human life. However, deliberative democrats must nevertheless agree to insist at the outset that democracy is the primary value to be upheld in the deliberations—that although participants in the debate may rely on and express strongly held religious beliefs, religious beliefs in and of themselves cannot and must not determine the course or the scope of the debate.

This might sound uncontroversial, but, in practice and in theory, it has important consequences that many of my fellow citizens (and many of my fellow humans) will not be willing to accept. In the course of this paper, for example, I will rehearse my position on prenatal screening for fetuses with disabilities, a position I set forth in chapter 2 of *Life As We Know It*, and I want to acknowledge immediately that there are many people of sound mind and good will who disagree with that position. Some of them, as I have learned over the past few years, dispute not only my calculations of the relative social goods and harms of prenatal testing, but also, and more stringently, my primary assumption that public access to and public uses of prenatal testing should be thought of in terms consonant with democratic values. As one of my interlocutors once put it, in the course of a 1998 discussion at Syracuse University's Center for Human Policy, it may be that democracy is not the highest value in such affairs. To people who start from the premise that there are divine laws, the meanings of which are not subject to human deliberation, and which therefore must be imposed on humanity by the representatives of the deity, nothing I say in this paper will have sufficient persuasive force. And this is but one example of a pervasive philosophical problem that some have called, for better or worse, "postmodern": when the claims of two systems of belief—democracy and theocracy, perhaps—are opposed, there may not be any way to reconcile them under the heading of a third, more universal term. We may instead be faced with a fundamental incommensurability between discourses of justification, such that we may not even have the language in which to agree to disagree. Understanding this problem, then, and admitting that I will have no common ground with some participants in debates over genomics even as I appeal to the practice of deliberative democracy, I will begin my discussion of biotechnology and disability by drawing on my training as a literary critic and theorist of popular culture—in this case, approaching intractable moral and political questions by talking about a Hollywood movie.

In January 2002, I attended a colloquium held at Washington University in St. Louis, the inaugural event for a semester-long discussion entitled "Law and Human Genome Project: Research, Medicine, and Commerce." The first plenary speaker, appropriately enough, was Dr. Francis Collins, Director of the National Human Genome Research Institute, and in the course of his paper—in which, incidentally, he called on Congress to pass a genetic nondiscrimination law—he briefly mentioned "the *Gattaca* scenario," that is, a dystopian world in which parents select for genetic traits among fertilized eggs, and in which job interviews, school placements, and indeed life prospects consist of blood tests and urine samples. The first thing to be said about the *Gattaca* scenario is that scientists tend to evoke it primarily to disparage or dismiss it: more often than not, the purpose of such rhetorical gambits is to counterpose modest, circumspect, prudent scientists (who know what can and can't be done when faced with chronic myeloid leukemia), to fuzzy-headed, antediluvian humanists who come into these discussions armed only with citations from Nathaniel Hawthorne and Aldous Huxley. This has been the dynamic for over a decade now: the humanists speak of dystopia and brave new worlds every time the scientists record an advance in genomics research, and the scientists dismiss fictional representations of brave new worlds as the products of the overheated imaginations of scientific illiterates on the wrong side of the two-cultures divide. And, needless to add, I choose my examples advisedly: chronic myeloid leukemia (CML) is a disease that now constitutes one of the success stories of the new genomics. People with CML seem to have a rare condition in which segments of chromosomes 9 and 22 fuse in such a way as to produce the abnormal protein BCR-ABL, and researchers have therefore developed a drug that blocks this protein and thereby kills leukemia cells. The drug is called Gleevec; it represents a new kind of cancer treatment that relies on what's called "molecularly targeted drugs," and it was approved by the FDA in May 2001. I am fully aware, in other words, that the genomics research that might give us *Gattaca* has already given us Gleevec. Nonetheless, I want to contest this rhetorical opposition between scientists and humanists, and I also want to contest the related rhetorical opposition between scientists and journalists—in which, again, scientists are prudent and circumspect whereas journalists are more or less wild-eyed snake-oil salespersons, ever ready and willing to declare on the covers of weekly newsmagazines that crime, disease, and even infidelity are all in our genes. When it comes to the new genomics, we will do well to be skeptical of such polar oppositions between pure research scientists and their various critics and champions. That's partly because the public understanding of claims about genetic screening will help to determine what kinds of genetic screening are requested (prenatal and preimplantation); but it's mostly because if you pay due attention to the history of eugenics, you will find that scientists themselves have been their own worst popularizers. And I mean "worst" in the sense that they have been purveyors of bad science, from Francis Galton right through to William Shockley and Arthur Jensen, and also in the sense that they have been purveyors of bad popularization, promising certainty where there is none to be had and offering human cloning by means of 1-800 numbers. So before I move to the world of *Gattaca*, I need first to challenge the standard opposition between, shall we say, science and popular mechanics when it comes to questions about improving the human stock.

The world of *Gattaca* is basically a world of advanced genetic screening, no more, no less: the technology the film envisions does not involve cloning, nor is there any suggestion of genetic therapies for "previously existing" conditions. Of course, in order for a society to attain the kind of knowledge of genetics evidenced in *Gattaca*, it would have to have done a great deal of very specific genomics research, some of which could conceivably require the cloning of human embryos for biomedical purposes. But this is not part of the explicit premise of the film. That premise is this: in the "not too distant future," prospective parents will reproduce by means of in vitro fertilization and genetic screening. Children conceived the old-fashioned way—indeed, in the back seat of a car, as the movie suggests—are called "faith births"; they are increasingly rare, and they make up the lower echelons of society. The movie's protagonist and narrator, Vincent, played by Ethan Hawke, is such a faith birth—excluded from school as a child because the school cannot afford the insurance costs he entails and consigned to a life of janitorial work even though he dreams of space flight, hoping against hope to work someday at the giant aerospace corporation, Gattaca. When Vincent is born, his initial genetic analysis reveals the following: he will have a 60 percent probability of a neurological condition, a 42 percent probability of manic depression, an 89 percent probability of attention deficit disorder, a 99 percent probability of heart disease, and a life expectancy of 30.2 years. His devastated parents decide to have a second child in vitro; "like most other parents of their day, they were determined that their next child would be brought into the world in what has become the natural way." Their consultation with the geneticist goes like this: he informs them that four extracted eggs have been fertilized—two boys, two girls—and that they have been fully screened for "critical predispositions to any of the *major* inheritable diseases." "You have specified," he goes on,

> hazel eyes, dark hair, [pause, smile] fair skin. I've taken the liberty of eradicating any potentially prejudicial conditions—premature baldness, myopia, alcoholism and addictive susceptibility, propensity for violence, obesity, etc.

Marie: We didn't want—I mean, diseases, yes, but—

Antonio: Right, we were just wondering if it's good to leave just a few things to, to chance.

Geneticist: You want to give your child the best possible start. Believe me, we have enough imperfection built in already. Your child doesn't need any additional burdens. And keep in mind, this child is still you—simply the best of you. You could conceive naturally 1,000 times and never get such a result.

The counselor smiles indulgently when he refers to the parents' preference for "fair skin," because, as it happens, he himself is black (he is played by Blair Underwood); by this we are led to understand that in *this* brave new world, genetic discrimination is common but racial discrimination a thing of the past. The point is reinforced a few minutes later, when Vincent recounts his experience as a young man seeking employment: "Of course," he tells us, "it's illegal to discriminate—genoism, it's called—but no one takes the law seriously. If you refuse to disclose, they can always take a sample from a door handle, or a handshake, even the saliva on your application form. If in doubt, a legal drug test can just as easily become an illegal peek at your future in the company." And we see a human resources manager wordlessly challenging Vincent to provide a urine sample. The manager, like the genetic counselor, is black.

I find this aspect of the movie's premise fascinating for two reasons—not only because the film so clearly (if counterintuitively and counterfactually) disarticulates racism from genoism, but also because it establishes *Gattaca* as, in part, a film about civil rights. It is, I might say, one of the few science fiction films that is centrally concerned with discrimination in employment. I do not say this facetiously: readers of Philip K. Dick will remember, for example, that one of the things that got lost in translation between the 1969 novel *Do Androids Dream of Electric Sheep?* and the 1982 film *Blade Runner* is the category of human "specials," people neurologically damaged by the nuclear fallout of world war and derisively referred to as "chickenheads" or, in severe cases, "antheads." The Voigt-Kampff empathy test, which we do see in the film and which bounty hunters known as "blade runners" employ to distinguish humans from androids trying to pass as human, was originally devised, Dick points out, to identify "specials" so that they could be sterilized and consigned to the lower echelons of society. Where *Blade Runner* obscures the relation between disability and employment (even placing the novel's "special," J. F. Sebastian, in the role of a high-ranking engineer for the Tyrell Corporation), *Gattaca* builds its plot around that relation. Moreover, the relation between race and disability is one of mutual implication in *Gattaca*: unable to pursue a career in aeronautical engineering because of his genetic makeup, Vincent decides to become a "borrowed ladder," using the bodily fluids and effluvia of Jerome (Jude Law) in order to obtain the clearance necessary for employment at Gattaca. Jerome is a former world-class athlete who was struck by a car; permanently disabled and visually marked by the most common sign for physical disability, his wheelchair. He literally sells his genetic identity to Vincent even as he himself descends into bitterness and alcoholism. In other words, *Gattaca* is not only the leading example of the science fiction employment discrimination genre; it is also a member of one of the oldest genres in African American fiction, the passing narrative.

Of course, it's possible to say that the film is simply wildly and bizarrely optimistic that a society so obsessed with genetics will be a society without racism. But I want to suggest that the film is wildly and bizarrely optimistic in another respect—in its assumption that a society so obsessed with genetics would have passed a law forbidding genetic discrimination, the very law our own Congress has so far failed to pass. (In the absence of such legislation, Burlington Northern Railroad conducted secret genetic tests on employees in 2000 and 2001, quite foolishly looking for a genetic marker for a predisposition to carpal tunnel syndrome. The railroad halted the tests when it was met with a federal lawsuit brought by the Equal Employment Opportunity Commission.) Look again at the kinds of genetic screening that the film presents as uncontroversial: all major inheritable diseases, as well as "potentially prejudicial conditions" such as premature baldness, myopia, alcoholism and addictive susceptibility, propensity for violence, and obesity. When Victor's mother objects, saying, "We didn't want—I mean, diseases, yes," surely the phrase she's thinking of and not uttering is something like "a designer baby." Which is, of course, precisely what she gets— still you, just the best of you. And the reason she goes along with it is that her child doesn't need any additional burdens—especially barriers to employment.

The *Gattaca* scenario presents a challenge especially to people like me, who have thus far combined political support for reproductive rights, a defense of technologies of prenatal screening, a critique of cost-benefit analyses of human worth, a stringent skepticism about the workings of our privatized and deeply inegalitarian insurance and health care system, and, last but not least, a defense of an aggressive social welfare state that provides needs-based benefits to children and adults with disabilities. My position on what is now called the ethics of selective abortion of fetuses with disabilities is a difficult and uneasy one. I agree almost entirely with the 1999 Hastings Center Report on the Disability Rights Critique of Prenatal Genetic Testing (see also Parens and Asch 2000), but I have to admit that the report cites *Life As We Know It* approvingly, so my agreement with it might sound uncomfortably like my agreement with its agreement with me. Be that as it may, I believe that the fetus does not have a moral status equivalent to that of a child unless and until it is viable outside the womb, and I support the right of prospective parents to terminate pregnancies even for reasons that I would regard as trivial or wrongheaded. Rayna Rapp's wonderful book, *Testing Women, Testing the Fetus* is replete with accounts of such parents, including those who told Rapp that "having a 'tard, that's a bummer for life" (91) or those who insisted that if the baby "can't grow up to have a shot at becoming the president, we don't want him" (92)—in regard to a fetus with Klinefelter's syndrome, on the basis of whose diagnosis the parents terminated the pregnancy. I remain unpersuaded that there are transcendent moral virtues to be advanced by compelling such parents to bear children with disabilities. For that reason, I have insisted that it is more consistent with the principles of democracy for people like me to *persuade* prospective parents and genetics counselors not to think of amniocentesis as part of a search-and-destroy mission, and to persuade them that many people with disabilities, even those disabilities detectable in utero (like Down syndrome), are capable of living lives that not only bring joy and wonder to those around them but are fulfilling and fascinating to the people living them as well. But I will not argue that some forms of childbirth should be made mandatory, nor will I demand that prospective parents be barred from obtaining genetic information about the fetus if they so desire such information.

At the moment, I submit, our society has achieved a shaky but substantial consensus that it is morally acceptable to screen fetuses for profoundly debilitating conditions such as Tay-Sachs disease, which involves severe and ceaseless suffering, but morally unacceptable to terminate a pregnancy solely with regard to gender. Everything else— Down syndrome, Huntington's disease, multiple sclerosis, leukemia— falls at various points in the nebulous area between, and thus far, we have apparently decided to leave decisions concerning such conditions up to the people who will be most affected by them. But precisely by taking the question of abortion out of the moral equation, the *Gattaca* scenario compels us to ask which "potentially prejudicial conditions" we would allow prospective parents to eliminate if the technology were available. That is to say, even in the not-too-distant-future, we might feel a profound moral repugnance at the idea of terminating a pregnancy simply on the grounds that the fetus has a genetic propensity for obesity, myopia, or premature baldness. But if we could select against these features at fertilization, would we do so, and what moral grounds would we offer for refusing to do so and preventing others, by law, from doing so?

My tentative answer runs as follows—and it is tentative, as I am still feeling my way through the moral thickets thrown up with each new advance in genomics. In a world that possesses the kind of genetic knowledge we envision in *Gattaca*, bioethicists, philosophers, presidential commissions, and humanists like me would have made the argument that while it is acceptable to screen for major inheritable diseases, screening out the "potentially prejudicial conditions" enumerated above would be highly controversial and far from routine. Our society would have had, and would still be engaged in, a wide-ranging debate about what kinds of disabilities do involve profound suffering or significantly diminished life chances for those who have them, and (by contrast) what varieties of human embodiment may be undesirable or inconvenient but, on the whole, do not constitute conditions so prejudicial as to jeopardize the life chances of those who have them. Such a debate would acknowledge, moreover, that many disabilities are not detectable genetically, and that no amount and no degree of prenatal screening or in vitro engineering will produce a world free of people with cerebral palsy or pneumonia, not to mention people who are hit by cars. And finally, such a debate would focus not only on potentially prejudicial conditions but on actually existing prejudices, extending the protection of the social welfare state to stigmatized populations while working also to destigmatize previously stigmatized identities (as had been done, evidently, with regard to people of African descent in the world of *Gattaca*). The debate would produce a boundary of the unacceptable, just as we now have agreement on the desirability of forbidding human cloning to produce children and a looser agreement on the permissibility but undesirability of selective abortion for gender. And in democratic fashion, the debate would seek to honor liberal freedoms and ideas of individual autonomy in decision making while insisting nonetheless, as both Mill and Rawls would tell us, that democracy does not have to honor all the preferences and desires of every person therein.

It pains me to admit it, but in such a society, the only people with Down syndrome would be those among the "faith births," those conceived without the benefits of in vitro fertilization and genetic screening. There would still be Deaf people in this society, partly because our social debate would have concluded that deafness is far less damaging to one's life prospects and one's ability to participate fully in the social and political life of the polity than is Down syndrome, and partly because some of our fellow citizens would have determined that Deaf culture is valuable, distinctive, and worth preserving in its own right, and that it has no linguistic or cultural counterpart among people with

Down syndrome. I say this not because I desire to see a world absent of people with Down syndrome; I cannot even imagine what it would be like to desire such a thing. But I imagine that if we had the power to screen for major disabilities, inheritable diseases, and potentially prejudicial conditions, many (but by no means all) of my fellow humans would see the elimination of Down syndrome as a social good, fewer would see the eradication of deafness as a good, and fewer still would see the eradication of myopia or baldness in those terms. Such would be, in my estimation, the results of a democratic deliberation about disability and genomics.

Yet one of the factors that makes this discussion so difficult and so tangled is the fact that, as I mentioned at the beginning of this paper, many of us do not agree that societies should decide such matters by democratic deliberation. One of the conundrums of such a deliberation, in other words, is that it necessarily includes the voices of those who do not believe that it should be taking place, or those who believe that it should take place but that it should not be morally binding or dispositive when it comes into conflict with moral absolutes. In a brief discussion of *Life As We Know It* for the Christian book review *Books and Culture*, for example, Jean Bethke Elshtain responded to my limited defense of abortion rights and prenatal testing by accusing me of "subtly but inexorably blowing out the moral lights among us, as Lincoln said of Douglas's defense of popular sovereignty in the matter of slavery." I find this remark at once exceptionally offensive and insufficiently morally serious, insofar as it rests on Elshtain's remarkably unelaborated claim that "the fetus, of course, is human all along—what else can it be?" For thinkers like Elshtain—and I assure you I am not overstating the political extremism at work in her essay—there can be no debate about the morality of prenatal screening and selective abortion, because there can be no debate about the moral status of the embryo: there is one correct position (that the embryo is fully human and deserves all the protections associated with ideas of inviolable human dignity), and then there are the positions of reprobates and infidels.

Nothing is more tedious, however, than an author complaining about a hostile (or, in this case, a hostile *and* dishonest) review,[1] so I will move on to a more consequential debate—the debate conducted and concluded in 2002 by the President's Council on Bioethics and the resultant report published under the title *Human Cloning and Human*

Dignity: An Ethical Inquiry. As was widely reported, the council unanimously endorsed a ban on so-called "reproductive cloning," which the Council called "cloning-to-produce-children," and, by a complicated 10-to-7 margin, a four-year moratorium on so-called "therapeutic cloning," which the council called "cloning-for-biomedical-research." As Leon Kass argued in the winter 2003 issue of *The Public Interest*, the Council construed the debate over cloning-for-biomedical-research as a clash between competing moral imperatives:

> On the one hand, we acknowledge that the research offers the prospect, though speculative at the moment, of gaining valuable knowledge and treatments for many diseases. On the other hand, this practice would require the exploitation and destruction of nascent human life created solely for the purpose of research.

This seems to me a reasonable statement of the conflict, even though the second sentence is almost a word-for-word repetition of Charles Krauthammer's partisan account of the debate in his personal statement in the report's Appendix. Yet within only a few sentences, Kass reframes it in a way that can only be called tendentious.

> Each side recognized that we must face up to the moral burden of either approving or disapproving this research, namely, on the one hand, that some who might be healed more rapidly might not be; and on the other hand, that we will become a society that creates and uses some human lives in the service of others.

Note what has happened between these two passages: both hands have changed substantially. On the one hand, at first, we had speculative research that offers the prospect of treating many diseases; but now, blocking that research means only that "some who might be healed more rapidly might not be," and not that "some who might be healed *will never be*." On the other hand, at first we were faced with the admittedly dicey prospect of "the exploitation and destruction of *nascent* human life;" but now if we permit cloning-for-biomedical-research, "we will become a society that creates and uses some human lives in the service of others," and there is no longer anything "nascent" about the embryo.

The reason the President's Council did not advocate an outright ban on cloning-for-biomedical-research is that it could not get a majority of its members, even among those who oppose abortion as a matter of principle, to agree that such cloning constituted an exploitation and destruction of human life in the sense that five-day-old embryos bear the same moral weight as do five-year-old children or fifty-year-old adults. James Q. Wilson, for instance, held to a position that would permit "biomedical research on cloned embryos provided the blastocyst is no more than fourteen days old and would not allow implantation in a uterus, human or animal." The council settled therefore for a moratorium that, as the report's executive summary puts it,

> provides time for further democratic deliberation about cloning-for-biomedical-research, a subject about which the nation is divided and where there remains great uncertainty. A national discourse on this subject has not yet taken place in full, and a moratorium, by making it impossible for either side to cling to the status-quo, would force both to make their full case before the public. By banning all cloning for a time, it allows us to seek moral consensus on whether or not we should cross a major moral boundary (creating nascent cloned human life solely for research) and prevents our crossing it without deliberate decision. It would afford time for scientific evidence, now sorely lacking, to be gathered—from animal models and other avenues of human research—that might give us a better sense of whether cloning-for-biomedical-research would work as promised, and whether other morally nonproblematic approaches might be available.

One objection to this statement, surely, is that it is question-begging, insofar as it presumes the very point that needs to be argued—namely, that cloning for biomedical research involves crossing a major moral boundary. This claim is precisely what many proponents of therapeutic cloning wish to contest. But the ancillary claim that biomedical research itself might benefit from this moratorium is simply disingenuous, as council member Elizabeth Blackburn pointed out in her personal statement in the appendix to the report: "It may sound tempting," Blackburn wrote, "to impose a moratorium to get more information, since, despite very promising results, it is true, at this early stage of the research, that we still know only a little. But that information can only be gained by performing the same research that the moratorium proposes to halt."

For those who grant the embryo full human status, however, there can be no justification for any cloning-for-biomedical-research; a moratorium would therefore seem to make little sense from that perspective as well, unless, of course, it is understood to be a temporizing move that will allow pro-life proponents to marshal their arguments and their social forces. Should that happen, and should we find ourselves living in a society whose democratic deliberations wind up in policy recommendations that ban both reproductive and therapeutic cloning, then we will have become a society in which the value placed on the moral status of the human embryo trumps all other considerations, including the moral injunction to alleviate plausibly remediable suffering. It is altogether possible to achieve such a society by democratic means, and I want to argue strenuously against it: for such a society would doubtless protect the lives and human dignity of people with disabilities, but at a terrible philosophical cost, that of enshrining essentially theocratic views about human life at the center of moral debate.

Having made clear my antipathy to theocratic conceptions of human life (even though I am duty-bound to respect them if they carry the day in democratic debate), I should immediately make it clear that I am by no means sanguine or unworried about the life prospects for people with disabilities in a society in which it is widely assumed that identifiable disabilities should either be cured or prevented. For the *Gattaca* scenario is just another vision of a world of eugenics, and the world of eugenics is already too much with us. On this count, contemporary professions of good faith among geneticists are no guarantee that they've learned the lessons of history. On the contrary, I would go further and argue, as Dorothy Roberts has done with great eloquence and power in *Killing the Black Body*, that the discourse of eugenics is not really "history" at all, and certainly not ancient history; public policy involving the involuntary sterilization of black women is as current as today's headline news. I am *not* saying, however, that the new eugenics is precisely the same as the old. I think the era of the Genome Project differs from the era of the 1929 Kansas State Fair in very important ways. You could call these "public" and "private" eugenics; or you could call them macro and micro eugenics, as Barbara Katz Rothman has done in *Genetic Maps and Human Imaginations*; or you could say that the old eugenics saw the human population as an aggregate of various ethnic and racial traits, some of which were not beneficial to the enlightened propagation of the species, whereas the new eugenics sees individuals as aggregates of biochemical traits, some of which are not beneficial to the families or nations in which they occur. I believe that this molecular view of the human is inadequate and incomplete, partly because genetics is an inexact science, a science of probabilities in which we cannot be sure how a biochemical predisposition may express itself, and partly because we have limited but conscious, self-reflexive control over how we express some of the traits we do express. It is one thing, in other words, to promote the Genome Project on the grounds that it will eradicate Tay-Sachs or Huntington's or Alzheimer's; I would regard each of these as a positive good comparable to the positive good of eradicating smallpox or tuberculosis. But it would be outrageous and destructive to think of individuals as agglomerations of traits like "propensity to become impatient," "facility with names," or "ability to memorize geographical maps," and to try to order an individual's genetic makeup accordingly.

And yet this is precisely what the libertarian opposition to the President's Council would have us do. Opposing bans on both "therapeutic" and "reproductive" cloning, the libertarian response to Francis Fukuyama's recent warnings about the hazards of biotechnology has claimed that the evil of eugenics as practiced in the first decades of the twentieth century was not that it was eugenics, but that the policies of involuntary sterilization and institutionalization represented forms of coercion visited on individuals by the state. "Private" eugenics, argue commentators such as Virginia Postrel, author of *The Future and Its Enemies*, is not an evil at all; it consists simply of individuals, in various familial arrangements, making rational decisions about what is best for them. In democratic deliberation over genetics and disability, therefore, I will want to argue against both Leon Kass and Virginia Postrel, the former because I believe he has an inadequate account of the social basis of moral debate and the latter because I believe she has an inadequate account of the social good. I will also argue against Postrel's position because I believe it is finally incoherent. For interestingly enough, when it comes to disability, suddenly even the most doctrinaire of libertarians resorts to arguing that some private decisions do not serve the public good. Postrel's reply to Fukuyama, for instance, draws the line when it comes to the Deaf lesbian couple who seek to have a Deaf child by means of in vitro fertilization. Postrel does not go so far as to argue that the state should actively prevent such births from occurring, but she does suggest that there is a compelling social interest in providing "disincentives" for people to bear children with disabilities: "Genetic intervention to create a deaf child," writes Postrel, "would constitute a form of child abuse that would *in theory* justify state action to protect the child." From this astonishing premise, she elaborates the following astonishing argument:

> So what to do about deaf parents who want deaf kids? I'm not sure. It's extremely dangerous to give state authorities power over reproductive decisions, and prohibiting parents from introducing genetically abusive traits in their children would require prenatal screening that could easily lead to mandatory eugenics. Ample historical experience, the kind conservatives generally value, tells us that it's wise to err on the side of preserving familial autonomy rather than looking for reasons to expand government regulation of family life. Respecting the family is a good general principle, even if in some cases (e.g., Andrea Yates) that respect has awful results.
>
> The best approach is probably an indirect one, such as some sort of liability for the doctors and others who perform prenatal genetic alterations. If the doctor who deliberately creates a deaf child has to pay for the youngster's special education, I don't think we'll see a lot of medically assisted child abuse. It would also help in the long run (though at the cost of considerable pain in the short run) to eliminate the many protections and privileges accorded disabled individuals.

The argument is astonishing in its obtuseness, and that it suggests (at the very least) that Postrel has not thought very long or very well about the position she espouses or about what she thinks of as "the many protections and privileges accorded disabled individuals." In fact, I thought it was so obtuse and ignorant an argument that I wrote to Postrel at her website in 2002, asking her what protections and privileges she might be thinking of. When she replied with some boilerplate about special education and medical services that are available to children with disabilities but not their disabled peers, I replied in turn, patiently pointing out that, for one thing, children with disabilities are eligible only for services that children without disabilities do not need. But what I really wanted to take issue with was Postrel's absurd cost-benefit analysis of human reproduction. Apparently, Postrel inhabits an imaginary world in which the lack of state provision for citizens with disabilities would create incentives for curing or preventing disabilities and disincentives for causing or maintaining disabilities. Such a world presumes, among other things, that we already live in a *Gattaca*-like world in which (a) people know in advance what disabilities their offspring might bear; (b) treatments in utero and in the germline can identify disabilities and diseases within hours of fertilization; (c) cerebral palsy, autism, pervasive developmental delay, and catastrophic illnesses can also be predicted shortly after fertilization happens; and (d) everyone is completely rational (but "rational" only in a sense that Postrel herself would approve of) about (a), (b), and (c), and made their reproductive choices accordingly. Speaking for myself—but, I hope, for some of my fellow humans as well—I would choose even to live in the world of *Gattaca* before I would choose to live in the world imagined by Virginia Postrel.

I draw from these recent debates two political paradoxes. The first is this: Many of the people who supported the passage of the Americans with Disabilities Act (ADA) of 1990 were, like White House counsel C. Boyden Gray, diehard antistatist conservatives, deeply opposed to gender-equity initiatives, race-based affirmative action, and civil rights laws generally. The reason that the ADA enjoyed such bipartisan support, however, was that its conservative and libertarian advocates championed it as a law that would free people with disabilities from dependence on the state. For them, the purpose of this public law was to return individuals with disabilities to the realm of the private. The second paradox is this: as our failure to pass genetic antidiscrimination laws indicates, there is no realm of the private. Disability is always and everywhere a public issue, a matter for public policy, even for political thinkers who otherwise have no conception of the public good. I want to suggest, then, that one way to think about disability, democracy, and genetics is to imagine that the public is not public enough and the private is not private enough. Those of us who support reproductive rights *and* a woman's right to prenatal testing *and* therapeutic but not reproductive cloning *and* the egalitarian provisions of the welfare state need to make the argument that intimate decisions about childbearing and care for people with disabilities need to be protected from state coercion yet supported by the state's apparatus of social welfare; at the same time, we need to make the argument that the state's apparatus of social welfare should seek to enhance the independence of people with disabilities from the state, but in order to do so, must recognize the very real dependencies associated with some disabilities, and must expand and enhance the roles of state-funded dependency workers. These are matters to be determined by democratic deliberation, a deliberation that must include the voices of people with disabilities and dependency workers. On one side of this deliberation, people like me will engage with moral absolutists for whom one value, one interpretation of human life, will always supersede all others; on the other side, we will engage with moral absolutists for whom one value, the value of individual freedom from state regulation, will supersede all others except when some individual decisions impose unacceptable social costs on other individuals. This, I submit, is the challenge of thinking democratically about disability in the age of genomics.

NOTE

1. I do not call Elshtain's review "dishonest" lightly. Two examples of its intellectual dishonesty will suffice. First:

> Bèrubè's [sic] cardboard cutout pro-life politician denies rights to living persons. One wonders who does this. Who are these people? He calls the implications of holding that humans have a right to life "only until they're born" staggering, and this would be true if anybody held to that view. But I can't think of a single pro-lifer who does, certainly not to judge from the literature I received from a number of pro-life groups.

I submit that Elshtain is too intelligent to intend this as a serious rebuttal. Take, for instance, the example of the obscure Texas pro-life politician who has not only executed dozens of living persons in his home state but has publicly mocked one such person's pleas for clemency. Is he the only pro-lifer in the country who believes that the state can take the life of a person?

Second, and far more seriously, Elshtain illegitimately interpolates her own words into a passage she cites from my book:

> If you had told me in August 1991—or, for that matter, after an amniocentesis in April 1991—that I'd have to feed my infant by dipping a small plastic tube in K-Y jelly and slipping it into his nose and down his pharynx into his teeny tummy, I'd have told you that I wasn't capable of caring for such a child. [In other words, had they had amniocentesis, they would likely have opted for abortion.] But by mid-October, I felt as if I had grown new limbs and new areas of the brain to direct them. (Elshtain's addition is in brackets.)

There are two things wrong with this interpolation. First, most children with Down syndrome do not require such care as this; therefore, amniocentesis would not have "told" us what we would need to feed Jamie with a gavage tube. Second, the bracketed sentence allows Elshtain to ignore my discussion of prenatal-care counseling and the provision of prospective parents with information about disabilities on pages 67–88 of my book. The reason *that's* important, in turn, is that my discussion of prenatal testing is targeted in large part at genetic counselors, whom I am energetically trying to persuade *not* to think of the detection of trisomy-21 as a search-and-destroy operation. And certainly it bears saying in so many words that I explicitly criticized the 90 percent abortion rate for fetuses with Down syndrome. Elshtain does not mention this at all.

As for Elshtain's suggestion that my position is analogous to that of Stephen Douglas with regard to slavery, all I can say is that had I pointed out that pro-life extremists habitually liken their opponents to the defenders of slavery, Elshtain would surely have scoffed at one of my "cardboard cutout" caricatures.

REFERENCES

Bérubé, Michael. 1998. *Life as we know it: A father, a family, and an exceptional child.* New York: Vintage.

Elshtain, Jean Bethke. 1998. "Idiots, Imbeciles, Cretins." *Books and Culture: A Christian Review* 4(1): 18.

Gattaca 1997. Directed by Andrew Niccol. Culver City, Calif.: Columbia Tristar Home Entertainment.

Kass, Leon. 2003. "Biotechnology: A house divided." *The Public Interest* 150 (Winter). Symposium with Diana Schaub, Charles Murray, William A. Galston, and J. Bottum.

Parens, Erik, and Adrienne Asch, eds. 2000. *Prenatal testing and disability rights.* Hastings Center Studies in Ethics. Washington, D.C.: Georgetown University Press.

Postrel, Virginia. 2003. The end of libertarians? http://www.dynamist.com/scene/apr29.html (accessed during week of April 29, 2003).

The President's Council on Bioethics. 2003. *Human cloning and human dignity: An ethical inquiry.* Washington, D.C.: Author. http://www.bioethics.gov/reports/cloningreport/fullreport.html.

Rapp, Rayna. 1999. *Testing women, testing the fetus: The social impact of amniocentesis in America.* New York: Routledge.

Roberts, Dorothy. 1997. *Killing the black body: Race, reproduction, and the meaning of liberty.* New York: Pantheon.

Rothman, Barbara Katz. 1998. *Genetic maps and human imgainations: The limits of science in understanding who we are.* New York: W. W. Norton.

CONTRIBUTORS

Kathleen S. Arnos
Professor
Department of Biology
Gallaudet University
Washington, D.C.

Karen B. Avraham
Associate Professor
Sackler School of Medicine
Tel Aviv University
Ramat Aviv, Tel Aviv

Michael Bérubé
Paterno Family Professor in
    Literature
English Department
Pennsylvania State University
University Park, Pennsylvania

Orit Dagan
Sackler School of Medicine
Tel Aviv University
Ramat Aviv, Tel Aviv

Brian H. Greenwald
Instructor
Department of Government
    and History
Gallaudet University
Washington, D.C.

Nora Groce
Associate Professor
School of Public Health
Yale University
New Haven, Connecticut

Shifra Kisch
Amsterdam School of Science
    Research
University of Amsterdam
Amsterdam, The Netherlands

Christopher Krentz
Assistant Professor
Department of English
University of Virginia
Charlottesville, Virginia

Louis Menand
Professor
Department of English and
    American Literature Language
Harvard University
Cambridge, Massachusetts

Anna Middleton
Genetic Counselor
Addenbrooke's Hospital
Cambridge, United Kingdom

Joseph J. Murray
Clinical Lecturer
Speech and Hearing Sciences
Indiana University
Bloomington, Indiana

Walter E. Nance
Professor and Chair
Department of Genetics
Medical College of Virginia
Richmond, Virginia

Arti Pandya
Assistant Professor
Department of Genetics
Medical College of Virginia
Richmond, Virginia

John S. Schuchman
Professor Emeritus
Department of Government
    and History
Gallaudet University
Washington, D.C.

Mark Willis
Research Coordinator
Office of Research Affairs
School of Medicine
Wright State University
Dayton, Ohio

INDEX

Page numbers in italics indicate figures or illustrations.

abortion: attitudes toward, 121–22; deliberative democracy and, 210–11; for hearing status, 139–41; prenatal diagnosis and, 128–29, 135–37, 208–9, 219n1

Abrams v. United States, 182

Abu-Loghod, A., 158

Abu-Shara Deaf community. *See* Bedouin Deaf community

age-related hearing loss, 81, 89

Alport syndrome, 83

American Breeders Association (ABA), 38–39

American Medical Association (AMA), 143

Americans with Disabilities Act (ADA) (1990), 217

amniocentesis, 139

anxiety gene, 8–9, 12–13

Armour, Robert, 47

Asad, Talal, 154

assortative mating, 98–101, 103

Authorization for the Destruction of Life Unworthy of Life (Hoche and Binding), 75

autonomy, 47–48, 102–3

autosomal dominant deafness, 85–86

autosomal recessive deafness: connexin 26 gene, 85, 116; gene frequency changes, *99*, 100; statistics, 97–98; in syndromic deafness, 113–15

Ballier, Wilhelm, 73

Bedouin Deaf community: about, 148–51; consanguinity, 151–52, 155–58; integrated signing community, 152–54; sociopolitical setting, 151–52, 158; stigma in, 161, 166–67; traditional vs. medical explanations, 149–51, 154–66, 168–69; women's issues, 156–58, 160–61, 163–64

Bell, Alexander Graham: on Deaf-Deaf marriages, 36–37, 42–47, 54–60, 101–2; eugenics and, 35–41; *Memoir Upon the Formation of a Deaf Variety of the Human Race*, 36–37

Bell, Buck v., 174–75, 182–83

Bell, Mabel Hubbard, 40, 70n80

Bengkala, Bali, 100

Berman, Marshall, 190

Bérubé, Michael, 208, 211

Biesold, Horst, 74, 76, 77n10

Binding, Karl, 75

bioethics, 211–15

biological determinism, 8–14

biotechnology, 211–15

Blackburn, Elizabeth, 213–14

Blade Runner, 207

Blank Slate, The (Pinker), 12

Boas, Frank, 15

Bourdieu, P., 163–64

Bowen, Catherine Drinker, 182

British Deaf Association (BDA), 143–44

Brown, Thomas, 57

Brunger, J.W., 138

Buck v. Bell, 174–75, 182–83

Burleigh, Michael, 75

Burlington Northern Railroad, 208

causes of deafness: connexin 26 gene, 85, 97–100, 111, 115–16, 120–21; connexin 30 gene, 116; Fay's data importance, 95–96, 103–4; genetic, 83–89, 111–16; *Khulf* (marking), 164–66; maternal fright, 58, 167, 171n18; nonsyndromic deafness,

83, 111–13, 115–16; syndromic deafness, 83, 112–15; traditional vs. medical explanations, 149–51, 154–66, 168–69

causes of disabilities, 26–27, 62n5

chronic myeloid leukemia (CML), 204

Clerc, Laurent, 48, 65n30

cloning, 211–15

CMV (cytomegalovirus), 96

cochlear implants, 89

Collins, Francis, 204

connexin 26 gene: about, 85, 97–100, 111; importance, 120–21; non-syndromic deafness, 115–16

connexin 30 gene, 116

consanguinity, 151–52, 155–58

cross-cultural studies: about, 23–25; Bengkala, Bali, 100; Martha's Vineyard, 29, 152–54; modernization and, 29–30; social beliefs, 25–29. *See also* Bedouin Deaf community

Crying Hands (Biesold), 74, 76, 77n10

cytomegalovirus (CMV), 96

Darwin, Charles, 2, 5–6, 14, 19–20

Davenport, Charles Benedict, 38–39

Davis, Lennard, 198

Deaf-Deaf marriages: A. G. Bell's view, 36–37, 42–47, 54–60, 101–2; connexin gene frequency and, 98–100; Deaf-hearing comparison, 51–52, 67n45, 68n52; eugenics and, 44, 60–61; prevention, 44–45, 47, 55–56, 63–64n17, 66n40; Sanders-Swett case study, 56–60; statistics, 48–50, 66n39; terminology, 45–46, 64n21; transnational dialogues, 42–53

Deaf education, 45–47, 76

Deaf educators, 43–44, 51, 53–56, 66n40

Deaf-hearing marriages, 51–52, 67n45, 68n52

deafness, causes of: connexin 26 gene, 85, 97–100, 111, 115–16, 120–21; connexin 30 gene, 116; Fay's data importance, 95–96, 103–4; genetic, 83–89, 111–16; *Khulf* (marking), 164–66; maternal fright, 58, 167, 171n18; nonsyndromic deafness, 83, 111–13, 115–16; syndromic deafness, 83, 112–15; traditional vs. medical explanations, 149–51, 154–66, 168–69

democracy, deliberative, 203, 209–11, 214–15, 217

"Democracy vs. the Melting Pot" (Kallen), 6–7, 14

determinism, genetic, 8–14

DFNB3 gene, 100

Dick, Philip, 207

dignity, 176–78, 183–84

disabilities: causes, 26–27, 62n5; modernization and, 29–30; social beliefs about, 29–30, 149–51, 154–56, 168–69. *See also* Americans with Disabilities Act

Do Androids Dream of Electric Sheep? (Dick), 207

Dougherty, George, 46, 52

Douglas, Mary, 162

Douglas, Stephen, 219n1

Down syndrome, 209, 210–11, 219n1

education, Deaf, 45–47, 76

educators of Deaf students, 43–44, 51, 53–56, 66n40

Elshtain, Jean Bethke, 211, 218–19n1

ethics: autonomy and, 102–3; cloning, 211–13, 215; genetic testing, 87–89, 141–42. *See also* abortion; eugenics

ethnicity, 6–8, 14

etiology of deafness. *See* causes of deafness

eugenics: A. G. Bell and, 35–41; British-America comparison, 63n11; Deaf-Deaf marriages and, 44, 60–61; in *Frankenstein,* 194–96, 199–200; in *Gattaca,* 196–97, 199–200, 205–8; genomic era and, 189–90, 205, 214–16; Nazi era and, 72–77, 175; oralism and, 39–40; proponents, 6, 36; sterilization, 2, 35, 64n17, 215

Everyone Here Spoke Sign Language (Groce), 2–3

evolutionary psychology, 8–14

Fay, Edward Allen, 49–50, 95–97, 103–4

Fischer, Eugen, 75

Five Factor Model, 12

FOXP2 gene, 100–1

Frankenstein (Shelley), 190–96, 199–200

Freeman, Matilda, 53

Friedlander, Henry, 73, 75–76

Fukuyama, Francis, 215–16
Future and Its Enemies, The (Postrel), 215–17

Gallaudet, Edward Miner, 48, 55
Gallaudet, Thomas Hopkins, 48, 65n30
Gallaudet Genetics Program, 118–20
Gattaca, 190, 196–200, 204–8, 210
gender issues: in Bedouin Deaf community, 156–58, 160–61, 163–64; public vs. private spheres, 52–53, 67n48, 67n50
gene frequency changes, *99,* 100–1
genes: for anxiety, 8–9, 12–13; connexin 26, 85, 97–100, 111, 115–16, 120–21; connexin 30, 116; *DFNB3,* 100; *FOXP2,* 100–1; *POU4F3,* 86
genetic counseling: in Bedouin Deaf community, 150; at Gallaudet, 118–19; nondirective approaches, 119, 143; translation of medical discourse, 156
genetic deafness: about, 111–13; autosomal dominant deafness, 85–86; causes, 83–89, 111–16; connexin 26 gene, 85, 97–100, 111, 115–16, 120–21; connexin 30 gene, 116; Fay's data importance, 95–96, 103–4; nonsyndromic, 83, 111–13, 115–16; syndromic, 83, 112–15. *See also* autosomal recessive deafness
genetic determinism, 8–14
genetic evaluation, 117–18
genetic mapping, 112
Genetic Maps and Human Imaginations (Rothman), 215

genetic nondiscrimination law, 204, 206, 208, 217
genetics research, 119–21
genetic testing: attitudes toward, 121, 131–40, 143–44; deliberative democracy and, 208–10; ethical issues, 87–89, 141–42; for hearing status, 142–44, 197, 216; implications, 116–18; preimplantation diagnosis, 141–42; prenatal diagnosis, 121–22, 128–29, 133–37, 143–44, 208–9, 219n1; as standard of care, 112
genoism, 197, 206–7
Gentile, Augustine, 96
Gillett, Phillip, 51, 54–55, 66n44
Gleevec, 204
Gray, C. Boyden, 217
Groce, Nora, 2–3

Harris, Judith, 12
Hastings Center Report, 208
Healey, George, 49
hereditary deafness. *See* genetic deafness
Hoche, Alfred, 75
Holmes, Oliver Wendell, 174–75, 182–83
"Holocaust Studies and the Deaf Community"(Friedlander), 75–76
Human Genome Project, 79, 81–83, 198
Huntington's disease, 209

immigration policies, 2, 6–8, 35
informed consent, 118, 177–84
intermarriage. *See* Deaf-Deaf marriages

James, William, 17–20
Jennings, Bruce, 175, 183
Jervell and Lange-Nielsen (JLN)
    syndrome, 114–15

Kallen, Horace, 6–7, 14
Kass, Leon, 212, 215–17
Kearney, Charles, 50–51
Khulf (marking), 164–66
Killing the Black Body (Roberts), 214
Kramer, Peter, 9
Krauthammer, Charles, 212

Lamarck, Jean Baptiste de, 5–6
Lane, Harlan, 152
legislation: Americans with
    Disabilities Act, 217; genetic
    nondiscrimination law, 204, 206,
    208, 217
Lenz, Fritz, 73
leukemia, 204, 209
Levine, Philip, 177
Lietz, Kurt, 76–77
Life As We Know It (Bérubé), 208, 211
Listening to Prozac (Kramer), 9
Locke, Alain, 14–17

Maginn, Francis, 52, 54
Marriages of the Deaf in America
    (Fay), 95
Martha's Vineyard, 29, 152
Martinez, A., 138
Mary Shelley's Frankenstein, 196
maternal fright, 58, 167, 171n18
mating, assortative, 98–101, 103
*Memoir Upon the Formation of a
    Deaf Variety of the Human Race*
    (Bell), 36–37
Menand, Louis, 2

Meniere's disease, 89
meningitis, 81
Metaphysical Club, The (Menand), 2
Mill, John Stuart, 20–21
modernization and disabilities, 29–30
Mossman, Mark, 194
Muhs, Jochen, 73
multiculturalism, 6–7
multiple sclerosis, 209
myosin VI gene, 86–87, *87, 88*

National Association of the Deaf
    (NAD), 73
National Deaf-Mute College, 46
natural selection, theory of, 2, 5–6,
    19–20, 98
Nazi era, 72–77, 77n7, 77n10, 175
noise-induced hearing loss, 89
nonsyndromic deafness, 83, 111–13,
    115–16
Not This Pig (Levine), 177
Nurture Assumption, The (Harris), 12

OCEAN personality spectra, 12
"On Nature" (Mill), 20–21
Operation T4, 75–76
oralism, 39–40, 45–47
Origins of Nazi Genocide, The
    (Friedlander), 73
otosclerosis, 89

Pendred syndrome, 81, 114
Pinker, Steven, 12
"Place of the School for the Deaf in
    the New Reich, The" (Lietz), 76–77
Postrel, Virginia, 215–17
POU4F3 gene, 86
Powers, Annabel, 50–51
pragmatism, 17–20

pregnancy termination. *See* abortion
preimplantation diagnosis, 141–42
prenatal diagnosis: abortion and,
    128–29, 135–37, 208–9, 219n1;
    attitudes toward, 121–22,
    133–34, 143–44
presbycusis, 81, 89
President's Council on Bioethics,
    211–13, 215
Proctor, Robert, 73
Prometheus, 191
psychology, evolutionary, 8–14

racial differences, 6–8, 14
Racial Hygiene (Proctor), 73
racism, 197, 206–7
Rapp, Rayna, 208–9
Roberts, Dorothy, 214
Ross, Edward, 6–7
Rothman, Barbara Katz, 215
rubella, 96

Sanders, Dorothy Bell, 59–60
Sanders, George, 56–60, 70n80
Sanders, Lucy, 56–60, 70n81,
    70n83, 70–71n84
Sanders, Margaret, 59–60
Sanders, Thomas, 56–58
selective mating, 98–101, 103
Shelley, Mary, 190–96
sickle-cell anemia, 102
signing communities, 29, 46, 100,
    152–54
social beliefs and disabilities, 29–30,
    149–51, 154–56, 168–69
speech development gene (*FOXP2*),
    100–1
Stargardt disease, 176

sterilization: of African American
    women, 214; *Buck v. Bell,*
    174–75; eugenics and, 2, 35,
    64n17; 215; Nazi era and,
    74–76, 77n7, 77n10
stigma, 161, 166–67, 210
Stoler, Ann Laura, 62n7
Swett, William, 57
syndromic deafness, 83, 112–15

Tay Sachs disease, 102, 209
teachers of Deaf students. *See*
    educators of Deaf students
Testing Women, Testing the Fetus
    (Rapp), 208–9
theory of natural selection, 2, 5–6,
    19–20, 98
Thompson, E. Synes, 51
Thorndike, Edward, 18
Turner, William Wolcott, 62n5
Turney, Jon, 191

Usher syndrome, 83, 113–14

Verkannte Menschen (Misjudged
    People), 72–73

Waardenburg syndrome (WS), 83,
    114
Weissmann, August, 5–6, 15
Wilson, James, 213
women's issues: in Bedouin Deaf
    community, 156–58, 160–61,
    163–64; public vs. private
    spheres, 52–53, 67n48, 67n50
worry gene, 8–9, 12–13

Yankee from Olympus (Bowen), 182